Space: The High Frontier in Perspective

Daniel Deudney

Worldwatch Paper 50
August 1982

Library of Congress Card Number 82-050920
ISBN 0-916468-49-6

Printed on recycled paper

Table of Contents

Introduction

Thales, the first of the ancient Greek philosophers, was a man whose knowledge of celestial occurrences was legendary in his time. Not much is known about him, but two stories about the effects of his fascination with the heavens capture both the promise and the peril of the space venture. One is that Thales, foreseeing the weather from the movement of heavenly bodies, was able to amass a fortune by shrewdly cornering the olive market. The other is that he fell into a well while looking at the sky. Today, as then, space can yield wondrous benefits, but it can also lead us to forget where we stand—thus inviting disaster.

Space is humanity's high frontier. Like all frontiers, space has produced unexpected treasures, generated strong enthusiasts, spawned wild speculations and been enshrouded in myth and false promise. Having spent several hundred billion dollars and a quarter of a century opening this frontier, it is time to assess if these efforts have helped solve some of the world's problems. Today space is the arena for both exciting exploration as well as economic exploitation. The many satellites orbiting the earth have a wide variety of functions, such as hurricane warning, observation of distant galaxies, business communications and crop forecasting. These and other space activities affect a surprisingly diverse and rapidly growing range of human concerns in ways that were scarcely imagined a few decades ago.[1]

Yet the space venture is today at a critical juncture. Sober realism as well as excitement and commitment must guide future efforts. The momentum of the long dominant U.S. civilian scientific space programs is slowing, and the superpowers are moving to put destructive weapons into space. The Soviet and American space monopoly is ending, and political conflicts surrounding commercially

I wish to thank Paige Tolbert for her assistance with the research for this paper, and Charles Chafer, Stephen Cheston, Leonard David, Thomas Karas, John Logsdon, Marcia Smith, Howard Symons, Adam Wasserman and David Webb for their reviews of the manuscript.

valuable space applications are emerging. Unfortunately, the growth of space's impact on vital human concerns has not been matched by an increased public awareness of the issues, effectively leaving these decisions to groups with agendas for war and national rivalry.

The quarter-century of activity in space has had its most significant impact on the strategic balance of power and the gathering momentum toward nuclear war. Military motivations led people to pioneer the high frontier and military activities remain the dominant use of space today. After rockets made it possible to hurl nuclear weapons across the oceans in a half hour, reconnaissance satellites eased tensions and aided arms control efforts. Today, the superpower cold war in space is entering a new, dangerous and destabilizing phase that—unless stopped—will make war more likely and less controllable. A resource that could be a decisive factor for world peace is in danger of contributing to humanity's last war.

In the struggle to protect the earth from overpopulation, ecological degradation and resource depletion, space has a great, largely unfulfilled role to play. It can be valuable not as a source of energy or materials, nor as a place to house the world's growing population, but rather as a tool to assist problem solving on earth. Orbiting satellites could be indispensable instruments in the scientific quest to understand how the planet's atmosphere, oceans and life forms interact, and how they react to the stresses being placed on them. Knowledge of these global systems is vital for a prudent response to such emerging global problems as carbon dioxide buildup, ozone depletion and deforestation. Satellites are already used to monitor forest and crop growth, weather patterns, mineral deposits and pollution sources. Unfortunately, these uses of space are not well funded and in some cases are even directly preempted by the military agenda.

Keeping space in perspective is difficult. Space is not a technology, a program or a cause—it is a place. Only a handful of human beings have ever been there, yet space activities affect the routine existence of most people. Space is only 80 miles from every person on earth—far closer than most people are to their own national capitals—but it is so vast that in it the earth is only an insignificant speck of dust. By asking how space activities affect life right here on earth, a sense of human perspective can be maintained.

7

Because space is so unlike anything experienced on earth, it is tempting to believe that earthly problems—resource conflict, pollution, scarcity—can somehow be left behind. Unfortunately humans have carried these problems with them. The most valuable areas of space near the earth are increasingly plagued by very earthlike, human conflicts and scarcities. Political, not technological, skills will be required to solve them. The inescapably global space environment provides a logical arena for cooperation among nations—which will be valuable both in its own right and as a prototype for terrestial coexistence.

Opening the High Frontier

Space begins where the earth's atmosphere trails off into a vacuum some 50 to 100 miles above ground. For human beings space is an overwhelmingly harsh and alien environment. Without the moderating effects of air the sunlight is unbearably intense and areas in shadow are frigid. If exposed to the vacuum of space, human bodies would explode. These voids around the earth are also washed by various forms of radiation that would be lethal to a person without extensive shielding. Only by devising ways to travel great speeds and withstand extreme environmental stresses have people been able to explore space. Once objects are out there, they can indefinitely maintain the 18,000 miles an hour speed that is needed to stay in space without the friction of the air slowing them down or burning them up. But leaving and reentering the earth's atmosphere is quite difficult. Gaining the speeds necessary to overcome the earth's gravity requires large amounts of energy. And reentering the atmosphere without burning up requires extraordinary materials capable of withstanding the metal-melting temperatures generated as objects fall through air.[2]

Humans have dreamed of traveling to celestial bodies for millenia. Isaac Newton first realized that an object traveling fast enough could—like the moon—achieve a balance between the pull of the earth and its own momentum to achieve permanent orbit around the earth. Not until the first decades of the twentieth century were the technical problems of building rockets to overcome the earth's gravity solved. The three undisputed fathers of rocket science—the foundation of all space exploration and use—are Russian mathematician Konstantin Tsiolkovsky, American physics professor Robert Goddard and German engineer Hermann Oberth. These men independently and almost simultaneously made a string of important technical advances and a wide range of remarkably accurate predictions.[3]

It was, however, in Germany that a government first committed resources to realize the scientists' visionary plans. Forbidden by the Treaty of Versailles to build artillery, the German military in the twenties supported rocket development at a time when it was neglected by the much larger military establishments of Great Britain, France, the Soviet Union and the United States. By World War II, the German scientists had built rockets four stories high that could travel several hundred miles. During the war Germany launched 4,000 V-2 rockets, each tipped with a ton of high explosives, in a strategically indecisive but psychologically numbing assault on British cities. After the war, both the Soviet Union and the United States, relying in part on captured German scientists and equipment, forged ahead with rocketry in order to build ballistic missiles capable of carrying nuclear weapons long distances. Fitted with lighter payloads to achieve the faster speeds needed to orbit, these military missiles became the basis of space exploration.[4]

The real beginning of the Space Age was the October 1957 Soviet launch of a 184-pound satellite into orbit. Sputnik electrified the world and was seen by many in the US as a "technological Pearl Harbor." Sputnik not only symbolized the emergence of the Soviet Union as a technologically advanced society, but demonstrated for the first time that the Soviet military had the means to deliver nuclear weapons to the United States. The US, which had been planning to launch a scientific satellite as part of the International Geophysical Year in 1957, put an object into orbit in January 1958, but only after an embarrassing explosion of a rocket on the launch pad. Manned flight began in 1961 with the orbital flight of cosmonaut Yuri Gagarin.[5]

In the 25 years since Sputnik, well over 90 percent of all space activities have been carried out by the two superpowers. From the modest beginnings in the late fifties, the space programs of both the United States and the Soviet Union burgeoned into major pro-

"The Soviet Union has taken an important lead in internationalizing access to space by sending Cuban, East European, French and Mongolian cosmonauts into space."

grams in military reconnaissance and support, space science, communications, lunar and planetary exploration, and remote sensing of the earth, as well as the manned orbital and lunar missions that have captured the most public attention. Today slightly over 4,000 artificial "moons" orbit the earth, with some 250 of them still operating. Space has become an arena of increasingly routine, commercial activities even as breathtaking discoveries continue.[6]

Prestige and national rivalry have fueled most of the civilian space efforts of both the US and the Soviet Union. Goaded by Khrushchev's boisterous proclamations of Soviet space superiority, President Kennedy pledged that the United States would land a man on the moon "within this decade." Although it unquestionably produced solid technical and scientific achievements, this type of prestige-motivated use of space spectaculars carried high costs. Lives have been lost and the benefits to science have been lower because of the "space race" mentality. According to Soviet space watcher James Oberg, Khrushchev's obsession with being first accounted for the largest space disaster—the death in 1960 of some of the country's leading scientists. Under strict orders to achieve a launch while the Soviet premier was at a U.N. function in New York, the rocket scientists abandoned the standard safety precaution of draining the fuel from a stalled rocket before closely inspecting it, and they were killed when the rocket suddenly exploded. The United States paid a lesser price, one of resource use, when in the interest of speed it decided to build a one-shot moon rocket rather than assembling one in earth orbit, an accomplishment that would have had continued relevance for a broad range of space activities.[7]

Without a doubt the Soviet Union has achieved more of the dramatic space "firsts" than the United States. (See Table 1.) Yet assessing the detailed accomplishments of the USSR is difficult due to the government's penchant for secrecy, the absence of publicly stated goals and the tendency to doctor historical evidence. Clearly, Soviet leaders have consistently placed a high priority on space activities, linking this exploration to the most important accomplishments and destinies of socialism. The Soviet Union has also taken an important lead in internationalizing access to space by sending Cuban, East European, French and Mongolian cosmonauts into orbit. The Soviets devote about 2 percent of their gross national product (GNP) to

Table 1: Space Milestones

Event	Year	Country
Artificial Satellite	1957	USSR
Rocket to Moon	1959	USSR
Interplanetary Rocket	1960	USSR
Man in Space	1961	USSR
Woman in Space	1963	USSR
Space Walk	1965	USSR
Man on Moon	1969	United States
Robot on Venus	1975	USSR
Robot on Mars	1976	United States
Probe to Jupiter and Saturn	1979	United States
Reusable Spacecraft	1981	United States

Source: Various news reports of the events.

space, while the United States spends 0.5 percent of its GNP. In absolute terms, the USSR probably now spends twice as much as the United States, about as much in real terms as the US spent at the peak of the Apollo moon effort in 1966.[8]

Despite a steadier and larger commitment of money to space, the Soviet effort has not translated into technological leadership because the general technological base the Soviets draw on—computers, materials, optics—is not as strong as in the United States. Like their industrial and agricultural sectors, the Soviet space program has had a hard time getting more performance from fewer physical inputs. This lag in miniaturization is reflected in the greater weight and number of Soviet launches needed to perform a given task, often cited erroneously in the United States as proof of Soviet superiority.[9]

Over the last decade the Soviets have taken steady steps toward the construction of a permanent earth-orbiting space station. They have maintained a more or less continuous human presence in space, routinely setting and then surpassing the record for endurance in orbit.

"Like their industrial and agricultural sectors, the Soviet space program has had a hard time getting more performance from fewer physical inputs."

From 1978 to 1981 the 20-ton Salyut 6 was occupied almost continuously by a two-person crew periodically serviced and replaced by additional cosmonauts. These steps toward a space station have probably cost the Soviets as much as the entire U.S. moon effort, although the scientific and prestige payoffs have been much lower. The Soviets will probably build a 12-person station soon. This goal will be easier to achieve if their long effort to construct a very large booster like the one used in the U.S. Apollo program succeeds.[10]

The U.S. space program has been less constant and more reactive but more technologically sophisticated than the Soviet Union's. Starting from a definite position of inferiority in the early sixties the United States rapidly pushed ahead, landing two men on the moon and returning them to earth in July 1969. The Apollo program was by far the largest single U.S. space effort, and represents the most technologically difficult and expensive human achievement in space. In addition, despite a lower overall expenditure the United States has made more practical use of space for communications, remote sensing and national defense. The U.S. planetary and lunar programs have contributed more to scientific knowledge than those of the Soviet Union due to the high volume and quality of photographs and data that American space probes send back to earth.[11]

Labeled a "moondoggle" by its critics, the $25 billion Apollo program seemed to be a particularly extravagant use of resources at a time of rising public concern about social conflict and environmental decay on earth. The ambitious proposals to land people on Mars at the cost of $100 billion and to build a permanent earth-orbiting space station never gained political support. Instead, the U.S. space program in the seventies centered around the effort to build a reusable spacecraft—the space shuttle—and was given much lower budgetary and political status than in the heady days of the "moon race." Even with its emphasis on lowering the cost of commercially viable activities in near space and the strong support of the military, the space-shuttle program has been barely able to garner enough support for its shoestring budget.[12]

The reusable space shuttle, first tested in 1981, was expected to accelerate the exploitation of space by reducing the cost of putting an object into orbit and allowing the repair or retrieval of orbiting satellites. However, due to funding delays and cost overruns it now appears that the shuttle will be only marginally cheaper than the new generation of expendable rockets it was intended to render obsolete. The four shuttles that comprise the fleet, built for a total cost of about $20 billion, are expected to fly about 300 times over the next 12 years. For the first time, passengers not trained to fly into space will be on board to perform scientific experiments. Although no plans have been announced, the logical next step for the United States will be the construction of a space operations center in orbit close to the earth or a space platform without astronauts in higher orbit to consolidate the many useful satellites already there.[13]

Because of the widespread use of American-made communication satellites and U.S. willingness to assist other countries in using data from remote sensing satellites, the rest of the world has benefited far more from U.S. than from Soviet space activities. Both launch satellites into orbit for other countries, although both maintain tight export controls on many of the key technologies used in space. Despite much rhetoric in favor of international cooperation in space exploration, joint U.S.-Soviet ventures and collaboration have been rare. Their most important cooperative venture in space, the 1975 rendezvous of a Soviet Soyuz and an American Apollo spacecraft, yielded few tangible benefits but was a valuable lesson in cooperation. Yet, in part because of the American perception that the linkup was a one-way technology transfer to the Soviets, cooperation has fallen off sharply. Several agreements for information exchange are expiring and, in retaliation for Soviet activities in Poland, the Reagan administration has announced its unwillingness to renew them.[14]

If the political prospects for superpower space cooperation have ebbed, the technological opportunities have probably never been greater. While the Apollo and the Soyuz crafts that linked in space were roughly parallel systems, the current centerpieces of the U.S. and Soviet programs—the space shuttle and the space station—are perfect complements. The shuttle is much more advanced technologically than the space station, but a cooperative research effort where a Soviet space station was serviced by an American shuttle would mean the transfer of no more manufacturing and material technologies than is already available to the Soviets in open literature. By eliminating the Soviet need for a shuttle and the U.S. need for

"The current centerpieces of the U.S. and Soviet programs—the space shuttle and the space station—are perfect complements."

space stations, both countries could save money and—more importantly—gain tension-reducing insights into each other's space activities. A minimum first step would be a trial rendezvous to test rescue procedures that could save lives in the event of some future space accident.[15]

Space in the eighties is no longer likely to be dominated by the superpowers. Vigorous European and Japanese efforts, as well as smaller Chinese and Indian programs, are emerging. After an abortive effort to build a cooperatively managed space program in the sixties, the Europeans have fashioned the European Space Agency (ESA), a viable parallel to NASA. From a fully equipped space port in French Guinea in South America, ESA has successfully tested a rocket system, the Ariane, that is giving NASA's space shuttle stiff competition in the market for commercial space launches. ESA's technological maturity shows in its use of the highly efficient, but technologically very demanding, liquid hydrogen and oxygen fuel system that is used by the United States. In contrast, the Soviets—despite repeated attempts—have not mastered this technology.[16]

Japan's entry into space is entirely civilian in orientation. Currently at least a decade behind the United States, Japan has embarked upon a 15-year, $14-billion effort to develop a broad range of hardware and services. Already the country has captured the major share of the market for ground stations to receive satellite signals, one of several currently profitable areas of space. The Japanese program aims to make cheaper and more durable versions of U.S. systems, and a small, much less expensive shuttle is contemplated. A recent study by the powerful Japanese Ministry of International Trade and Industry (MITI) estimates that by the mid-nineties, the Japanese space industry will be a $4.5 billion a year business—as big as the present Japanese radio and television manufacturing industry. By the end of the eighties ESA and the Japanese space agency NASDA may be launching as many satellites as the US or the Soviet Union does, as they fight to capture significant commercial markets in launch vehicles, communications, science and remote sensing.[17]

Several Third World countries have also joined the "space club" by launching their own satellites. While many countries own satellites that have been launched for them by the superpowers, and even more own earth stations to receive information from satellites, the threshold for national space independence is a rocket capable of putting an object into orbit. China and India have built and successfully launched their own rockets and Pakistan is on the verge of doing so. These programs will probably not be strong competitors in the commercial markets for space services, but they give these developing countries both a civilian and a military foothold in space. China aims to use space for many of the military activities that the US and USSR have long performed there.[18]

In a development with widespread military and commercial implications, private entrepreneurs have begun building rocket launchers too. A German firm, OTRAG, rented a part of Zaire twice the size of Switzerland and set out in 1977 to build cheap rockets capable of putting objects into orbit. A series of unsuccessful tests, coupled with intense pressure from Soviet diplomats fearful of clandestine German military operations, forced OTRAG to move to Libya. There, much to the embarrassment of the German Government, which cancelled the company's export license, an OTRAG team is believed to have helped build missiles for Colonel Qaddaffi until its withdrawal late in 1981. A second private launch effort, Space Services, Inc., is testing rockets off the coast of Texas under tight U.S. Government regulatory controls. The firm's vice-president, Charles Chafer, estimates that simply adopting business rather than bureaucratic practices can give the company's Conestoga rocket a 15 percent cost edge over NASA's shuttle. Yet another group of U.S. investors has reportedly offered to pay at least $1 billion for a space shuttle.[19]

Within a decade, the United States and the Soviet Union may find they have borne all costs of opening the high frontier—and that others are profitably exploiting it. Having more countries focusing primarily on the civilian uses of space is a welcome counterbalance to the growing military emphasis of the superpowers, which has emerged as a major drag on civilian space activities. Instead of serving as a source of spin-offs for civilian progress, the military is increasingly preempting technologies with valuable civilian applications. A recent study of U.S. space activities by the Office of Technology Assessment concluded that the nation would not be a strong competitor in emerging space markets unless NASA's charter is altered to allow the agency to operate as well as develop systems and unless clear goals for the program are set forth. The Reagan administration is betting that the private sector will move in to fill the void, setting the stage for another high-stakes race between Japanese and European government-supported enterprises and purely private American firms.[20]

One of the benefits of space exploration that is difficult to quantify is the spin-offs—technologies discovered in the process of exploring space that have found application in unrelated human activities on earth. These range in significance from freeze-dried orange juice and heat-resistant materials to sophisticated sensors. U.S. military scientists trying to put computers into the cramped space of a rocket nose cone turned semiconductors from a scientific oddity into the basis of a multibillion dollar industry that is profoundly altering many spheres of life on earth. Similarly, photovoltaic cells, a technology that may someday dramatically enhance the world's energy prospects, were first developed for commercial use by U.S. scientists to supply satellites with electric power. Foil and plastic laminates and strong resilient sealants have also been put to use in energy conservation efforts and pump design. NASA has taken a deep interest in such spin-offs, hoping to justify further space expenditures.[21]

The discovery of very valuable technologies during space research programs does not mean, of course, that they would not exist otherwise. Yet there is no denying that having to adapt to the peculiar demands of the space environment has led to ingenious new technologies. The spin-offs from space programs typically use fewer physical inputs to yield higher performance. In a world that must wring more out of less, the lessons to be learned in the space environment could be particularly valuable ones.

War or Peace in Space

The desire to gain military advantage originally drove people to use outer space. The first use of space—to rain bombs on distant countries—created an environment of unprecedented tension and insecurity. Then reconnaissance satellites reduced tensions and made verifiable arms control agreements possible. Now, however, the military use of space has entered a third phase. Whereas the military programs of the sixties and the seventies used space for information and communication, the superpowers have begun a race—for the first time—to place weapons of destruction in orbit. Unchecked, these developments could largely cancel out the positive benefits of space, divert funds from civilian programs with great scientific and commercial potential and increase the likelihood of thermonuclear war.

Since their almost simultaneous creation during World War II, the technologies of atomic fission and rocket propulsion have been closely linked. Missiles and rockets differ only in where they are aimed and what they carry. For the last 30 years, the balance of power between the United States and the USSR has centered on the inventories of nuclear-tipped missles. Perhaps the most telling—but least discussed—indication of the motivation behind acquiring space launch capability is the almost perfect correspondence between the "nuclear club" and the "space club." Of the six nations that have built space launchers, only Japan does not also possess nuclear explosives. The two most recent entrants in the space club—India and China—are the last two nations to explode nuclear weapons. All indications are that Pakistan is next in line to both explode a nuclear bomb and build its own space launcher. And it is surely no accident that Libya's well-known quest for atomic weapons was paralleled by a less well-publicized effort to lure German missile scientists to that country.[22]

The military activities of the superpowers have been and continue to be the dominant use of space. The U.S. Congressional Research Service estimated in 1981 that the Soviet Union has successfully launched 858 military and 392 civilian missions, while the United States has sent up 420 military and 327 civilian payloads. These numbers are only approximations because some flights are dual-purpose and because the Soviet Union reveals little about the intent of its launches. As in overall space budgets, the Soviet lead in military missions is not a reflection of a menacing "space gap" but of the longer lives of U.S. satellites. In the late seventies the U.S. military space budget finally surpassed the civilian; the military's share is still on the rise due to diminished spending for civilian programs and the breakdown of détente. Counting the share of the space shuttle's cost assignable to military missions, 20 to 27 percent of the budget of NASA (the government's civilian agency) may really

"The superpowers have begun a race—for the first time—to place weapons of destruction in orbit."

be for military purposes, so the military accounts for close to 75 percent of U.S. space spending. (See Figure 1.) Similar budgetary trends—less visible because of the organizational unity of the Soviet space program—are occurring in the USSR. The U.S. Department of Defense estimates that 70 percent of Soviet space spending is directly military in nature, with a further 15 percent of combined civilian and military purpose.[23]

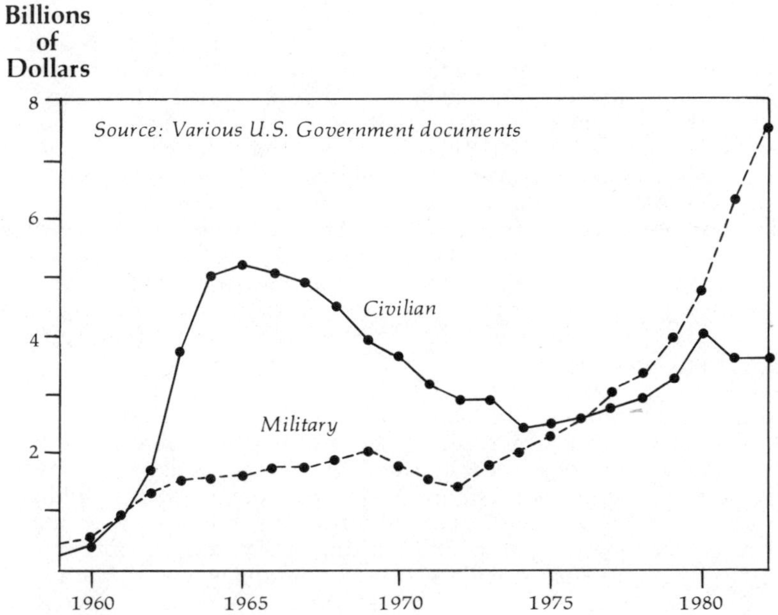

Figure 1: U.S. Civilian and Military Space Funding, 1959-1982
(with shuttle budget equally divided between military and civilian)

Cloaking military programs or the development of a technology with high military potential under the veil of the peaceful use of outer space is a common practice of the superpowers. The centerpieces of their civilian space programs in the seventies—the space shuttle and the Salyut stations—serve important military functions and would probably not have been funded had it not been for military value. In the Salyut stations, cosmonauts carry out both photo reconnaissance and surveillance. The space shuttle, redesigned to meet military needs, was funded due to the military's political clout and is likely to serve the military more than any other user.[24]

By far the most beneficial military use of space is for reconnaissance and surveillance. Since the early sixties the United States and the Soviet Union have used orbiting cameras to observe in detail military activities deep within each other's territory. Since as a closed society the Soviet Union is able to keep its military activities secret, these missions are of greatest benefit to the United States. In 1961 satellite photographs of Soviet missiles and bomber facilities exploded the idea of a U.S.-Soviet "missile gap" and averted an expensive U.S. crash catch-up program. After viewing satellite photos one leading Senate advocate of the "missile gap" theory admitted he had been 99.9 percent inaccurate. Several years later, President Johnson claimed that the $34 billion to $40 billion spent on the space program had saved ten times as much by reducing arms expenditures. For the first time since the atomic age began, an advance in technology helped to constrain the rapid spiral of destructive capability.[25]

During the seventies, the value of military space reconnaissance and surveillance to peacekeeping grew still further. Abandoning their early claim that observation satellites were "spies" committing "espionage," the Soviet Union began to build systems similar to those of the United States. The superpowers felt secure enough to negotiate and sign agreements limiting the numbers of strategic weapons. Without such satellites to verify compliance with the treaties, it is extremely doubtful that either country would have entered into these pacts. Reflecting the importance of the satellites, the SALT I Treaty explicitly outlaws attempts to interfere with "national technical means of verification."[26].

Satellites have further enhanced global political stability by enabling leaders to monitor crises, watch for remote nuclear weapons tests and communicate with each other quickly. To monitor compliance with the Nuclear Non-Proliferation Treaty of 1970 and Limited Test

"In 1961 satellite photographs of Soviet missiles and bomber facilities exploded the idea of a U.S.-Soviet 'missile gap' and averted an expensive U.S. crash catch-up program."

Ban Treaty of 1963, the United States has placed observation satellites in high, large vista orbits to scan continuously the remote regions of the earth as well as near space for nuclear detonations. In 1977, the Soviet Union spotted a nuclear test facility under construction in a remote area of South Africa. After intense diplomatic pressure, South Africa dismantled the test site. This early satellite warning of an impending test is widely credited with averting a profoundly destabilizing addition to the "nuclear club." Reconnaissance satellites have been used to monitor regional conflicts that could draw the superpowers into military confrontation with each other. Intense reconnaissance satellite activity was reported during the 1971 Indian-Pakistani war, the 1967 and 1973 Arab-Israeli wars and 1980 turmoil in Iran. Linked by the satellite-based "hot line," U.S. and Soviet leaders can be in direct contact with each other in a matter of minutes—which may someday prevent an ambiguous situation from turning into a war. By reducing mistrust based on misinformation and by allowing leaders to communicate as rapidly as they can retaliate, satellites have added a much-needed element of stability to the volatile international political situation.[27]

The exact capabilities of U.S. and Soviet reconnaissance and surveillance satellites are, of course, tightly guarded state secrets. Yet from a variety of anecdotal evidence and from the steps the U.S. Air Force took to design MX shelters that could deceive Soviet satellites, it is clear that the observational power of these systems is awesome. The oft-reported claim that these satellites can read a license plate is probably not far from the truth. The backbone of the U.S. "sky spy" system since the early seventies has been "Big Bird," a satellite 50 feet long and 10 feet in diameter, weighing 10 tons. From its 100-mile high orbit, Big Bird can scan every spot on earth in daylight every other day, sending a constant stream of TV images and periodically dropping canisters of high-resolution film into the atmosphere where they are recovered by specially designed aircraft as they parachute to the ground.[28]

To broaden the use and availability of surveillance satellites for peacekeeping and treaty monitoring, Howard Kurtz, founder of War Control Planners, Inc., has promoted for a decade the idea of establishing a network of satellites run by the United Nations. This idea gained official support in 1978 when France proposed an International Satellite Monitoring Agency (ISMA). As outlined by then President Giscard d'Estaing before the U.N. General Assembly, ISMA would extend the benefits of surveillance satellites to nations without space capability, would permit the Security Council to monitor crises and border disputes and would lay the groundwork for monitoring compliance with the treaties banning chemical and biological warfare and environmental modification. Depending on whether ISMA obtained technology from the superpowers, a basic monitoring system would cost between $1 billion and $2 billion a year, more than the entire U.N. budget. The United States has strongly opposed ISMA, arguing that sensitive issues of data interpretation would be impossible for an agency operating by majority rule. Support for ISMA has been strong, of course, among the many countries that are not likely to soon have their own sophisticated observational satellites. The first question to resolve is who would have access to the data generated by the satellites.[29]

The hard-line opposition of the United States to ISMA—quietly shared by the Soviet Union—reflects a desire to continue monopolizing the political advantage of satellite technology. This opposition however, may be shortsighted since the US stands to lose most from a breakdown of the "open skies." Unless its benefits are more widely dispersed, this technology could come to be widely seen as a form of illegitimate spying, perhaps leading to a treaty banning or limiting such systems. Any number of countries might be willing and eventually able to enforce such a ban. Although ISMA's opponents protest its cost, if satellite verification of regional arms control efforts is half as successful as it has been between the superpowers, the agency could pay for itself many times over in reduced arms expenditures.[30]

Despite the positive contributions of space reconnaissance and surveillance to arms control, the military communication, navigation and scientific satellites have begun to unsettle the balance between the superpowers, undercutting many of the security accomplishments of the SALT treaties. Most destabilizing has been the use of satellite data to calibrate ballistic missile trajectories. Indeed, the so-called "window of vulnerability" motivating the U.S. arms buildup would not have occurred without geodetic satellites that precisely measure anomalies in the earth's gravitational field. With flight paths adjusted to compensate for these imperfections that previously would have

"The so-called 'window of vulnerability' motivating the U.S. arms buildup would not have occurred without geodetic satellites."

drawn them off course, Soviet and American missiles are so accurate that they can blow up missiles in concrete silos reinforced with hardened steel. This improved accuracy has given an edge to the side that strikes first. To ensure that their land-based missiles are not lost in a sudden attack, both superpowers have adopted quick-reaction command systems that effectively put a hairline trigger on nuclear war. This in turn greatly raises the chances of accidental nuclear war.[31]

Other systems on the drawing board or just becoming operational will erode security further, tempt preemptive first strikes and undercut arms control agreements. In a move yet unmatched by the US, the Soviet Union has a series of satellites that scan the oceans with high-powered radar pulses to locate large U.S. naval ships. Linked to air- and sea-based missiles, these satellites could give Soviet commanders the information they need to launch a surprise attack on U.S. ships. The new U.S. NAVSTAR global positioning network of 18 satellites, to be fully operational in 1988, will make the location of submarines so precise that the missiles they are carrying could knock out hardened missile silos, adding a new mutual "window of vulnerability." Another U.S. system in the works, IONDS, will record detonations of U.S. nuclear warheads, allowing commanders to use fewer weapons to destroy with assurance a given target. IBM physicist Richard Garwin estimates that IONDS will multiply the effectiveness of U.S. missiles by 40 percent, enabling the existing Minutemen III ICBM's to destroy as much as the proposed MX system.[32]

All these new systems undercut arms control accomplishments by multiplying the effectiveness of remaining weapons. Because many of these satellites probably would not work well for very long after the outbreak of a nuclear war, due to the aftereffects of the bomb blasts, they have little deterrence value and in fact encourage the belief that a surprise nuclear attack could really work. Controlling these qualitative improvements through arms control agreements is virtually impossible—scientific information about things such as gravitational anomalies is hard to ban once acquired. And many of the force multipliers, such as communications and navigational satellites, are inseparable from increasingly useful civilian space systems.[33]

The growing importance of satellites in fighting wars has spawned both Soviet and U.S. anti-satellite weapons. Whereas the military space programs of the sixties and the seventies provided surveillance information and communication channels, the new generation of satellite killers involves for the first time the launching of destructive systems into space. If unchecked by appropriate arms control measures this technology could largely cancel out the peaceful benefits of space while opening up an expensive and volatile new dimension to the arms race.[34]

The first anti-satellite system, employing nuclear warheads, was built by the United States during the sixties, extensively tested, and then dismantled. The United States gave up this approach when it became clear that a nuclear blast in space would not only obliterate its target but send out pulses of electromagnetic radiation so strong that all U.S. satellites not shielded by the earth at the time of the blast would also be destroyed. This ability of a few large nuclear blasts in near space to clear the skies of operating satellites is always within easy reach of both the superpowers. Simply by sending up several of their larger warheads and missiles either country could even the score from any unfavorable encounter in space.[35]

Since 1968, the Soviet Union has tested a non-nuclear anti-satellite system that is more discriminating in its destructiveness. In 20 tests thus far, a satellite hovers near the intended victim and then explodes, shattering the fragile satellite with a shower of shrapnel. In June 1982 the Soviet Union for the first time tested this satellite killer in conjunction with large-scale ballistic missile test launches from silos and submarines. This coordinated action feeds fears in the United States that the Soviet Union is preparing a first strike. These tests are, however, indistinguishable from Soviet preparations to shoot down American targeting satellites in the event of U.S. first-strike attack.[36]

Despite the wide attention given the Soviet satellite killer in the Western press, the system involves no new technological breakthroughs. Indeed, any country capable of a routine orbital rendezvous has within its grasp a satellite killer equally as sophisticated as the Soviet's. The current Soviet anti-satellite system is of minimum threat to the United States program because most U.S. military satellites are much higher than the Soviet system reaches. Controlling these satellite killers would be difficult, at any rate. With orbital maneuvering so common and the presence of an explosive device so difficult to detect, an agreement outlawing the technology could be verified only through close examination of every object put into orbit. Still, a ban on further tests would block the evolution of this system into a more comprehensive and threatening weapon.[37]

The U.S. military, however, has designed and will soon begin testing a new satellite killer that, if deployed, would leapfrog over the Soviet capability. While achieving the same end—destruction of an orbiting satellite—the U.S. system employs a small homing missile that collides with the intended target at high speed after being launched from a high-flying fighter plane. As with the Soviet system, this satellite killer can reach only targets in low orbit. From the standpoint of arms control and crisis management, the U.S. system will be far more destabilizing. The Soviet orbital rendezvous system can be tracked as it is launched and as it closes in on its target, but the U.S. direct-ascent system could strike with little warning from fighter planes located anywhere in the world. Short of using a nuclear blast, for the USSR to sweep U.S. satellites from the skies, many visible launchings of large rockets over several days would be required, but a globally coordinated U.S. strike could probably destroy most operable Soviet satellites within a day. More than just a catch-up effort, the U.S. system is a technological advance into a whole new plateau of danger. The direct-ascent satellite killer also threatens a high percentage of Soviet military satellites because most of them are in low orbit.[38]

Sporadic negotiations between the USSR and the United States to control anti-satellite weapons during the late seventies were broken off by the United States to protest the Soviet invasion of Afghanistan. During the talks one stumbling block had been the Soviets' insistence that the U.S. space shuttle be defined as an anti-satellite system. The USSR claims that the shuttle's ability to maneuver in orbit and to bring objects back to earth means the United States can destroy or hijack Soviet satellites. In response, the US claims the shuttles are too expensive ($1 billion apiece), too scarce (only four will be built), and too fragile to use as space fighters, especially since the Soviets could quite easily mine their satellites to explode if tampered with. Although the shuttle is itself not a very plausible anti-satellite weapon, its versatility does give the United States an important ability to fight in space. The U.S. Air Force's strong continuing interest in the shuttle adds to these Soviet fears, stimulating Soviet tests that could threaten U.S. space assets.[39]

Time is running out to control these new weapons. From the standpoint of verifiable arms control, the testing—not the deployment—of satellite killers is the realistic point of no return. Once thoroughly tested, these systems will be an arms control verification nightmare since many orbiting Soviet vehicles or many U.S. fighter planes could harbor satellite killers. The recent decision by the United States to accelerate testing of the direct ascent homing missile but to delay anti-satellite negotiations has been justified as a "bargaining chip" strategy, to develop a weapon to be negotiated away at some future date. Testing this satellite killer might restore "balance" in the short term but would open up a volatile, destabilizing new area of arms competition. The United States depends far more than the Soviet Union on space for both civilian and military purposes, giving the government a powerful motivation to check the spread of these satellite killers. Unilateral U.S. restraint and an immediate resumption of negotiations would do more than further tests to enhance the security of both nations.[40]

Beyond these simple adaptations of routine technologies to destroy satellites, Soviet and American military scientists are enthusiastically working to build space-based energy ray weapons. No longer just the stuff of Buck Rogers fiction, laser and particle beam systems are being developed with hundreds of millions of dollars in research funds and already exist in prototype. At a time of general budget cutting, in 1982 the U.S. Congress approved a funding increase for this technology even though the Department of Defense insisted that no more money could be absorbed productively. In various tests, lasers placed on mountaintops, on ships and in airplanes have shot down oncoming missiles. The inability of such weapons to penetrate fog, smoke, dust or rain makes them even more promising in the airless voids of space. The earliest uses of space-based lasers will probably be to blind surveillance satellites.[41]

As envisioned by U.S. planners, powerful lasers in orbit would shoot down intercontinental ballistic missiles as they arc briefly into space. Due to the scattering effect of clouds and air, these lasers would have little effect on objects on the ground. A space ray system that could

"Once thoroughly tested, these systems will be an arms control verification nightmare."

shoot down the existing arsenal of Soviet ICBMs would cost—conservatively—between a phenomenal $500 billion and $1 trillion, and would place in orbit over 300 satellites larger than any yet built. An effective space-based ballistic missile defense system has one very appealing quality—it could eliminate the possibility of nuclear missle attack against population centers. Proponents argue that such a weapon would be inherently defensive in character, thus ending rather than fueling the nuclear arms race.[42]

The ultimate defensive weapon—the dream of military planners from the time of the Great Wall to the Maginot Line—is not, however, any more likely to be realized in space than on earth. The new exotic space weapons would merely extend into yet another realm the same stalemate of forces present on earth—at exorbitant cost.

Using space-based lasers as a ballistic missile defense suffers from the same problems—in spades—that led the US and USSR to agree by treaty to abandon all anti-ballistic missiles a decade ago. The central problem of all anti-ballistic missile systems is the ease with which the highly sensitive radars that track oncoming missiles can be utterly blinded. The massive aftershock of electronic radiation from a single nuclear explosion in space or in the upper atmosphere would disable sensitive circuits, radar screens and computers and make it impossible to locate warheads moving at bullet speeds. The systems' other militarily fatal flaws include their ability to be countered by decoys, mirrored warhead surfaces, spinning warheads and warheads laminated with burn-resistant surfaces. Even using optimistic cost figures it is clear that a dollar of defense expenditure could be effectively countered for a dime—a defense bargain no nation could afford.[43]

The momentum of present military developments in space bodes a bleak future out there—and down here on earth. At a time of growing resource competition and renewed ideological hostility, space militarization is being aggressively promoted by the Soviet and U.S. military establishments, undermining important security interests in each nation. Unilateral restraint, bilateral treaties and multilateral initiatives should be high national security priorities for both the United States and the Soviet Union. Other countries just now beginning to use space could play an active role in prodding the superpowers into restraint. Otherwise, space will slip from the realm of peace into one of war.

26 Learning about Earth from Space

The most valuable contribution space activities have made is the information they have provided about the universe and about the earth. From a lofty vantage point, scientists are learning answers to some critical life-and-death questions about the earth—how the atmosphere, oceans, sunlight and life forms interact to make the planet habitable and how human actions are altering those systems. Beyond invaluable lessons about the mechanics of "spaceship earth," space activities are emerging as vital tools in the management of terrestial systems and resources.

Despite these vital contributions, science and resource monitoring have been the poor cousins of the war machine in space. An early space science pioneer, Dr. James Van Allen, discovered the radiation belts around the earth that today bear his name after he convinced the military to substitute his instruments for the bags of sand they were using in test missiles to simulate the weight of warheads. With meager, declining budgets and hand-me-down technology from the military, these earth-keeping space activities are as neglected on the high frontier as on the earth itself. Nor was the largest civilian space venture—the Apollo moon project—a particular priority for scientists: a 1963 poll of several hundred U.S. scientists found almost no support for the project. Today, the momentum of space science and resource monitoring is faltering at a time when global monitoring systems should be established to keep tabs on the faltering health of the earth.[44]

Thus far the biggest scientific advances from space exploration have been in astronomy and related sciences. The scientific investigation of space has centered on sending probes, and, in the case of the moon, people, to other parts of the solar system. Probes landed on Venus and Mars by the US and USSR gave scientists firsthand information about the surface and atmosphere of these planets, and provided a strong indication that they do not harbor life. And the recent U.S. Pioneer and Voyager missions transmitted thousands of spec-

"Machines, not humans, have been and will continue to be the principal explorers of the solar system."

tacular high-quality close-up photographs of Jupiter and Saturn and their many moons.[45]

The American lunar landings represented the high point of human journeying into space, not likely to be surpassed for many decades to come. Machines, not humans, have been and will continue to be the principal explorers of the solar system. The instruments sent to the moon were of more scientific importance than the astronaut's reports. Indeed the Soviet Union's Lunokhold traveled 50 kilometers over the moon's surface, and returned with 330 grams of lunar rock—without humans ever leaving the earth. The mission to land people on Mars proposed in the US during the early seventies would have cost more than $100 billion, yet two Viking probes capable of performing a broad array of experiments without astronauts were sent for $500 million. As robots become even cheaper and more versatile in the years ahead, the justification for sending people into space could diminish still further.[46]

There are many other objects in the solar system—notably asteroids, comets and the moons of Jupiter and Saturn—that scientists want closer looks at, but planetary probes, which cost from $200 million to $1 billion, are increasingly hard to finance. Budget cuts in the U.S. program mean that the pace of solar system exploration for this decade will be considerably slower than in the last and will virtually grind to a halt in the nineties unless funding commitments for new missions are made by the US or other governments in the next few years. The budget of the U.S. planetary science program has been cut so deeply in the last year that expensive probes already far from the earth will be permanently turned off. Dr. Eugene Levy, Chairman of the Planetary Sciences Board of the U.S. National Academy of Sciences, estimates that $400 million—not the current $150 million—is needed each year for the US to maintain a healthy program.[47]

As these deep-space voyages taper off, a series of powerful observation systems that will be placed in earth's orbit but trained on distant objects will be the center of scientific attention. The most exciting of these will go into operation in 1985 when the U.S. shuttle places into orbit a telescope 2.5 meters in diameter. Although there are larger telescopes on mountaintops on earth, their observational powers are limited by the distorting effects of the atmosphere. The space telescope will have ten times better resolution than ground-based telescopes, and will be able to detect objects 50 times fainter than at present, thus expanding the observable volume of space 350 times. Many astronomers believe this new window on the universe will revolutionize their science, allowing, for example, the detection and observation of planets around other stars.[48]

Spending money to explore the universe has sparked controversy between those who believe more pressing needs exist on earth and those who seek knowledge. Unlike pure science on earth that often yields some practical benefit later, astronomical discoveries are a particularly pure (that is, useless) form of knowledge. Space explorations have, however, spawned two new sciences—comparative planetology and exobiology, the study of life elsewhere in the universe. Practitioners of these nascent sciences argue that learning how other planets evolved will yield insights into the evolution and fate of the earth. Pointing out that scientific knowledge is a group of generalities from many case studies, comparative planetologists and exobiologists question whether the knowledge of a planet—the earth—can ever be firmly scientific when it is based on an exhaustive study of only one example. Thus the hellish surface temperature of Venus, the result of an atmosphere of heat trapping carbon dioxide, is a partial case study of what the earth could be like if the "greenhouse effect" runs wild. The storm clouds on the fast-rotating planet Jupiter give scientists information on the effects of rotation speeds on weather patterns unobtainable on earth.[49]

Beyond such intriguing tidbits, comparative planetology has yet to transform human understanding of the earth, perhaps because scientists need to look more closely at the earth's sister planets, or perhaps because those planets are not very much like earth. On the theoretical side, however, James Lovelock, the inventor of a trace-element monitoring technology widely used in pollution assessment, came up with a systematic reinterpretation of life's role in the evolution of the earth—the Gaia Hypothesis—while designing the life detection experiments for the Mars Viking Lander spacecraft. (The Gaia hypothesis holds that life has transformed the earth to suit its needs rather than adapting to the earth.) By thinking about how to detect life on earth just by analyzing its atmosphere, Lovelock was led to a reconceptualization of life's interaction with its physical environment that is open-

"Observation satellites could be as important to oceanography, climatology and geology as the microscope was to microbiology, or the telescope to astronomy."

ing important new avenues of research in ecology, paleontology and pollution control.[50]

Space's biggest boost to knowledge of humanity's home may come from looking more clearly and directly at the earth itself, rather than at other, vaguely similar planets. Progress in sciences of the earth as a whole—oceanography, climatology and geology—is intimately linked with human concerns such as weather patterns, fisheries potential, pollution assessment, earthquake warning and mineral exploration. For these, observation satellites are vital because the objects of their study—the oceans, the atmosphere and the solid earth—operate on a planetary scale. Before satellites, scientists had to be content with scattered observations from ships, weather balloons and aerial photographs. Now for the first time they can actually see what they are studying. One major hurdle for all the whole earth sciences has been assimilating the sheer volume of satellite data—often thousands of times more than from traditional sources—into existing models and theories. Ultimately, observation satellites could be as important to oceanography, climatology and geology as the microscope was to microbiology, or the telescope to astronomy.[52]

Of the three earth sciences, climatology and geology have been most affected thus far. Satellites have been indispensable in monitoring the thin upper layers of the atmosphere and the trace gases such as ozone that shield the earth's surface from lethal solar radiation. Recently a satellite monitoring the amount of solar radiation striking the upper atmosphere found fluctuations of 0.2 percent in less than a week, leading scientists to think that changes in the sun itself may be a factor in climate change and weather patterns on earth. Mapmakers and geologists have used orbiting instruments to accurately map the earth and to measure its magnetic and gravitational anomalies, giving new insight into the planet's interior composition. Because microwaves and lasers between the earth and satellites can measure distances of thousands of miles with an accuracy of a few inches, geologists are for the first time able to track the slow push and pull of the earth's crust that cause earthquakes. The use of satellites in oceanography has lagged because the one ocean-monitoring mission, the U.S. SEASAT, went dead after only a hundred days in space. Another limitation on oceanography's use of information from space is the inability of satellites to measure temperature beneath the ocean's surface.[52]

The ability to look continuously at broad areas of the earth's surface is of practical as well as scientific value. Monitoring the earth for everything from weather forecasting and iceberg tracking to crop assessment and mineral exploration is known as "remote sensing." Remote sensing satellites are similar to those used by planetary scientists or military intelligence officials in basic technology and aim, but different in the type, resolution and volume of data they generate. Some remote sensing uses have become commercially important, others are poised to become so, and still others are in the research phase.

The first and still most frequent use of remote sensing in terrestrial resource management is weather satellites. From far above the earth's surface, cameras and other sensors provide meteorologists with broader pictures of weather movements than ever before available. Combined with the analysis from high-speed computers, meteorological satellites have made weather forecasting much less of a guessing game: today's 24-hour forecasts have the same accuracy—84 percent—as 12-hour forecasts did 15 years ago. With better prediction of severe storms, such as hurricanes and typhoons, evacuation warnings can be issued and lives saved. Since satellites began keeping track of hurricanes in the mid-sixties, no one has died because of deficient warning. Hurricane Camille, the worst storm of the century, caused minimal loss of life in 1969, whereas 1,500 people died in hurricanes in Mexico in 1959 and 5,000 in Texas in 1900. A cooperative typhoon-warning system being set up in East Asia should reduce the area's yearly storm damage of more than $3 billion. The Philippines, annually hit by four or five typhoons boiling suddenly off the Pacific, will be a major beneficiary. Within 15 years, global satellite imagery should enable meteorologists to make five-day forecasts that are as accurate as 24-hour ones today, which would translate into $5.5 billion of savings in agriculture and aviation in the United States alone.[53]

Beyond weather satellites, a wide array of remote sensing satellites stand poised to move from the research and development phase to routine daily use. Beginning in the early seventies, the United States and the Soviet Union put into orbit general purpose remote sensing satellites. France and Japan are expected to orbit civilian remote sensing systems within the next five years. Known as LANDSATS, these satellites have proved useful to farmers, foresters, shippers, highway builders, coastal zone managers and mapmakers.[54]

Remote sensing of living systems—crops, forests, grasslands, plankton and fisheries—could provide solid trend information on a truly global scale as well as having many uses in day-to-day resource management. The U.S. Department of Agriculture has used LANDSAT images of foodgrowing regions to improve crop harvests. Using similar data, U.S. officials in recent years have made more accurate preliminary estimates of Soviet wheat harvests than the Soviets themselves. Economist Klauss Heiss estimated in 1979 that the direct economic benefits of satellite crop forecasting have been about $325 million a year, far exceeding the $80 million cost of the program. In the United States, forest-products firms with large landholdings have found that satellite images provide not only broader coverage than ground-based assessments, but also a more accurate view of tree health than direct visual inspection. Thus far these systems are of value only over areas with large fields and low crop variety.[55]

Mineral and petroleum explorers have also benefited from remote sensing. By studying satellite images of known mineral deposits and then looking for similar formations elsewhere, geologists have been able to locate commercially valuable deposits. Most notable thus far has been the discovery of copper deposits in a remote region of Pakistan. How widespread such discoveries are likely to be is a matter of controversy among geologists. The oil companies have made extensive use of LANDSAT data, but the impact this has had on oil discoveries is hard to assess because the firms maintain a tight lid of proprietary secrecy on exploration techniques. Another useful application of remote sensing will be to monitor the reclamation of strip-mined land.[56]

Remote sensing satellites have provided particularly useful information about developing countries, if only because conventional mapping, mineral surveys and crop inventories have been so sparse. The United States has gone to great lengths to involve Third World countries in its LANDSAT program, heavily subsidizing the export of receiving stations and providing free training to scientists. Some seven developing countries now have receiving stations. Brazil, the second largest user of LANDSAT photos after the US, is assessing for the first time the amount and rate of deforestation over large areas of remote terrain. Brazilian scientists have replaced previous widely varying estimates with firm scientific ones and confirmed deforestation is occurring more rapidly than government officials believed. The satellite photos will also help Brazil target reforestation areas and monitor success. In the barren regions of the Middle East, LANDSAT images are used to track the patterns of moisture in remote desert areas, giving farmers warning of desert locust outbreaks.[57]

The prospects for greater routine use of LANDSAT-type satellites for resource management are today clouded by a series of institutional, political and financial problems. Because the data obtained by remote sensing satellites has commercial use, developing countries fear that multinational corporations geared to use the information will gain greater leverage on their economies and enhanced control over their resources. Several Third World countries, led by Indonesia, believe that no remote sensing data of a country should be acquired or released without the observed country's explicit permission.

The United States, supported by most of the other OECD members, maintains a policy of open access. In fact the United States sells LANDSAT photographs to anyone willing to pay the price without even asking for the purchaser's identity. A compromise position advanced by the Soviet Union would require the sensed nation to give prior consent for dissemination of images with greater than 50 meters resolution. The current U.S. LANDSAT satellites have a resolution of 80 meters, but a U.S. experimental system has a 30-meter resolution and the new French commercial system will go down to 10 or 20 meters. Stopping at 50 meters would eliminate some potentially valuable applications, such as detecting specific pollution sources, but going to 10 or 20 meters would generate information of military significance. The line between spying and science has all but disappeared with these technologies.[58]

Before a LANDSAT system can be routinely used the difficult task of matching users to information must be solved. Questions of what type of sensors are used, what resolution of images are available, how rapidly information is available and how frequently images are made of a given area would evoke a different answer from each potential class of users. The solution to this problem favored by the

"**Maps of the moon are better than those of the earth because military authorities allow lunar orbiters to carry more advanced cameras than those on civilian spacecraft that orbit the earth.**"

most active LANDSAT country, the United States, and its potential commercial competitor, France, is to turn the systems into profit-making operations. The United States has been the leader in developing and demonstrating LANDSAT technology, but the U.S. Government has no plans to follow the current experimental systems with an operational one that potential users would need to justify the costs of receiving and processing equipment. The official U.S. policy is that private investors should build and operate LANDSATs. France, however, is moving to fill this gap between the existing users' needs and NASA's experimental work with an operational system known as SPOT. Neither the quasi-government SPOT nor the strictly private system desired by the U.S. Government can operate profitably, however, just by selling images of France or the United States and international areas like the oceans, so they would be forced to make and market images of other countries.[59]

Turning LANDSATs over to private enterprise may, however, create as many problems as it solves. International pressure to restrict the systems may well increase if developing countries feel they are being extensively monitored for the benefit of the highest bidder. And scientists and resource managers in the Third World question whether LANDSATs designed with profit in mind will produce the kinds and quantities of information they need. A solution to these problems may be to develop international cooperative institutions that are run in a business-like fashion but that serve a broader agenda of scientific research and are more responsive to the needs and fears of the poorer members of the world economy.[60]

The future of both planetary science and remote sensing from space will to a great degree be determined by scientists' access to military technology. Although not all military reconnaissance and surveillance technology is of immediate use to science and earth resource management, the civilian programs could certainly benefit immensely from the transfer of some military technology. Dr. Farouk El-Baz, Research Director of the Center for Earth and Planetary Sciences at the U.S. National Air and Space Museum notes, for example, that maps of the moon are better than those of the earth because military authorities allow lunar orbiters to carry more advanced cameras than those on civilian spacecraft that orbit the earth. Much of the early earth-sensing technology was pioneered by the military and transferred to civilian uses later. But the military also prohibits the use of key sensing technologies by civilian groups. Thus the U.S. military forced the ill-fated SEASAT system to carry radar systems considerably below the state of the art, much to the dismay of some scientists whose experiments could have benefited from the more advanced radars. Point-source pollution monitoring will also suffer without the use of high resolution sensors now monopolized by the military. In the U.S. space effort, this military preemption of technology puts the country one step behind others in the commercial space race: the French SPOT, for example, will use sensors with roughly twice the resolution of LANDSAT, not because of French technological innovation, but because the U.S. military makes NASA use low-resolution sensors for fear of revealing secret technology.[61]

The Satellite Communications Revolution

Bold prophesies of future technologies are commonplace these days, but few have been as successful at this art as the British scientist Arthur C. Clarke, author of *2001: A Space Odyssey*. In 1945, Clarke published a detailed plan to use orbiting satellites for conveying messages from place to place on the earth, although he certainly did not expect realization of his idea within his lifetime. Yet today he regularly converses with scientists all over the world via satellite from his home in Sri Lanka. Communications satellites and their support systems are a rapidly growing industry because they enable people to transfer information more cheaply and reliably than ever before. Remote sensing applications may be plagued with mismatches between producers and users of data, conflicts over access to the product, and military monopolization of technology, but the communication satellite industry has integrated itself into terrestrial systems and pioneered highly successful international cooperative organizations that have thus far avoided paralyzing political conflict.[62]

The commercial use of space is dominated by communication satellites, which account for about 90 percent of the commercial benefits of space. Since Telstar, the first commercial communications satellite, started beaming signals in 1962, there has been an explosive growth in this industry. In 1965 less than 100 trans-Atlantic satellite tele-

"Leasing a standard telephone circuit for a year cost $32,000 in 1969, but only $4,680 now."

phone circuits were in use; in 1980 there were 20,000 connecting all parts of the globe. The number of circuits is expected to double by 1984, and to reach 130,000 by 1993. This tremendous growth has been fueled by an equally phenomenal fall in the price. Leasing a standard telephone circuit for a year cost $32,000 in 1969, but only $4,680 now. (See Figure 2.) Worldwide, satellite communications represent nearly a $2-billion-a-year business, a figure that is expected to reach $10 billion a year by the end of the decade.[63]

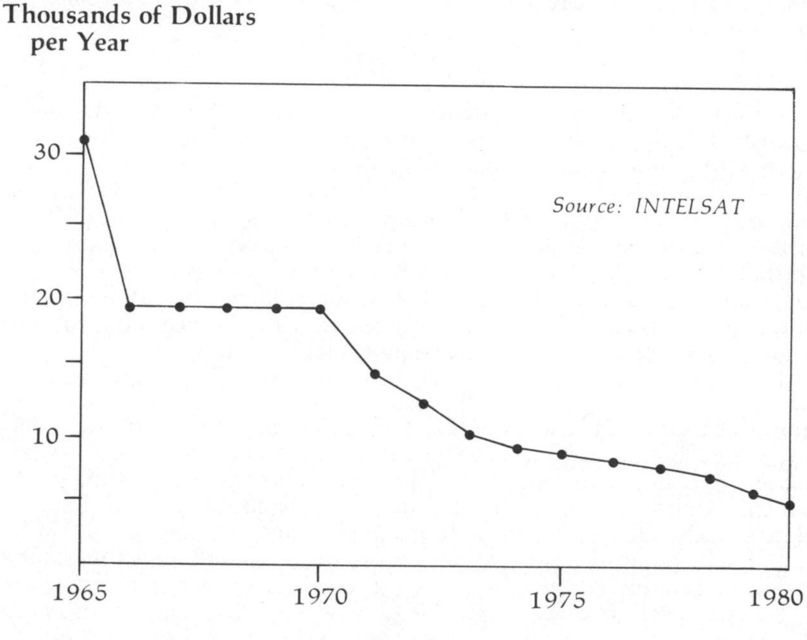

Figure 2: INTELSAT Unit Service Charge (in current dollars), 1965-1980

These satellites have created for the first time in history a genuinely global communications network. Twenty years ago Argentina and Chile talked to each other via New York exchanges, while Nigeria and Ghana spoke via London. At the core of this new planet-spanning communications system are the 15 satellites owned and operated by the International Telecommunications Satellite Organization (INTELSAT). Founded in 1964, INTELSAT now has 106 member nations. It is an outgrowth of—and until recently dominated by—COMSAT, the U.S. domestic satellite corporation. Created by treaties between countries but operated like a profit-making commercial enterprise, INTELSAT is the most successful cooperative international space venture. The Soviet Union built a separate network, INTERSPUTNIK, to provide services for Eastern Europe and its own vast reaches.[64]

While these satellites allow businesses in developed countries to lower the cost of telephone services, they offer developing countries a chance to fashion communication links where none existed before. Only the US, USSR and Canada had domestic satellite systems in 1975; by 1980 there were 15 domestic satellite systems in the Third World. This greater relative importance is dramatically reflected in the fact that Third World countries contain only 7 percent of the telephones in the world but account for 38 percent of the satellite communications traffic. Developing countries are, in essence, able to skip the costly stage of building long-distance telephone lines.[65]

Communications satellites are becoming larger, more complex and more expensive while receiving stations on earth become smaller, cheaper and more mobile. The first experimental communications satellite, ECHO I, was nothing but an inflated metal foil balloon that bounced radio waves back to earth. In contrast, the new INTELSAT VI will be a telephone exchange in space, as tall as a four-story house and capable of simultaneously receiving, processing and routing 37,000 telephone conversations. The resulting opportunity to miniaturize earthside receivers has contributed to a rapid proliferation of dish antennas. In the United States the number of receiving stations—typically owned by a business or government agency with a high volume of telephone use—has jumped from 50 to 3,000 in the last three years. NASA has even produced designs of a giant satellite linking a global network of personal mobile telephones that would be small enough to wear like a wristwatch.[66]

The ability of satellites to transmit information cheaply has a considerable effect on international commerce and politics. Thus far, it has led to a notable centralization of decision-making. Because the United States uses more satellite communications than any other country, the impact of the new technology on diplomacy, business and the armed forces has been most pronounced there. Ambassadors who previously had a degree of discretion in running the day-to-day contacts between governments are now effectively under the detailed control of the State Department officials in Washington. Similarly, large multinational corporations can keep much tighter rein on their far-flung operations from their central headquarters in London, New York, Singapore or Tokyo. Centralized control of distant military operations has also greatly increased. In the 1975 Mayaguez incident and the 1980 attempt to rescue hostages in Iran the President of the United States was in more direct control of military operations on the other side of the world than commanders a few miles from the battle have been in previous years.[67]

Another information service of growing economic importance—navigation satellites—is owned and operated by the military but used by thousands of civilians. Like remote sensing, naviation satellites face growing problems of civilian access to military technologies. The U.S. Navy's Transit satellite system, operational since 1963, provides signals that when processed by small computers reveal a ship's location to within 50 meters. Receivers, costing about $20,000 each, are used by an estimated 6,000 ships now that the U.S. Navy allows civilians access to the system. An even more advanced military network of 18 NAVSTAR satellites will, when fully operational in 1988, provide moving aircraft with highly accurate locational fixes. Present plans are to allow some channels, but not all, to be received by civilians. The receivers, no larger than a portable radio, are small and light enough to use in small private aircraft, battlefield units or cruise missiles.[68]

Satellites also transmit television signals, and this may be the use that has had the most immediate impact on the greatest number of people to date. News broadcasting has been the most dramatically affected. Appropriately, the first event watched live by people worldwide was the landing of the men on the moon, images of which were beamed from U.S. satellites rushed into place for the event. Live television broadcasts of sports events like the soccer World Cup matches and the Olympics are seen by as many as two billion people. Satellite TV broadcasts are bringing cultural interactions that for the first time begin to match the scale of the world's economic and political interdependencies.[69]

Television broadcasting via satellite has been slower to develop than telephone services and more controversial, reflecting in part its power to undercut politically sensitive and economically lucrative monopolies held by established TV networks. In the United States, local cable TV, two-thirds of which is transmitted by satellite, made the first inroads on the monopolies, and was followed in the early eighties by a spate of new networks that beam images directly to small antenna dishes. Already between 30,000 and 40,000 are in use and the number grows by 2,500 a month. In Western Europe, where almost all television broadcasting is done by government-controlled companies, Luxembourg recently licensed a private firm to begin satellite broadcasts. Because the country is so small, the broadcasts will spill over into large areas of neighboring countries, creating politically controversial competition for government stations and a lucrative new market for private advertisers. Fearing the intrusion of foreign TV broadcasts into its tightly controlled information market, the Soviet Union, joined by several Third World countries, has led an international effort to ban direct TV broadcast across national borders.[70]

Some planners in the Third World are enthused about satellite television broadcasting as both a means to communicate with large segments of the population out of the social mainstream and a way to assist in literacy campaigns. By cheaply bringing educational programs or entertainment to the countryside, satellites can help remove one of the incentives for urban migration. Studies in El Salvador indicate that villages with one or more publicly accessible televisions lose their population much less rapidly than those without them. In 1976, in a joint U.S.-India effort, direct-broadcast satellites beamed teacher education, student instruction, family planning, hygiene and agricultural programs to an estimated five million people in six different clusters of 2,400 Indian villages. Viewing the experiment a success, India has called the launch of a permanent, Indian-built system a top priority for its nascent space program. The biggest constraint on reaping the educational advantages of direct broadcast sat-

"Satellite TV broadcasts are bringing cultural interactions that for the first time begin to match the sacle of the world's economical and political interdependencies."

ellites is not the hardware—the satellites, the transmitters and the receivers—but the "software," such as programming geared to villages, teachers capable of integrating the images into meaningful lessons and financial support.[71]

Space Industrialization: The Mirage of Abundance

No one looking at space a quarter century ago foresaw all ways in which space systems have altered human life, and no one today predicting space activities over the next 50 to 100 years is likely to predict all the uses to which satellites will be put. Yet enough is known about the potential of space and the needs of earth that some ideas—large-scale space colonies, solar power satellites and asteroid mining—can be eliminated as practical and desirable goals. A modest industry producing certain high quality products is possible within this time horizon, but is unlikely to surpass the economic contribution of communications satellites. The longer term prospects for human ventures in space are, of course, not knowable. But large-scale space industrialization is not a viable solution to the pressing population, energy or resource problems of the earth.

The success of the satellite communications industry has spawned a great deal of speculation about the possibility of building industries in space. Experiments aboard the U.S. Skylab in 1974 revealed that the weightless, airless conditions of space enabled the production of certain goods that are impossible to manufacture on earth. Without gravity, crystals form much more regularly, permitting the creation of glasses and electrical devices that have vastly higher performance than any others. And the products can be formed without containers, eliminating impurities that are the major limit to the performance of certain optical and metallurgical products on earth.[72]

The drug industry could be a major user of weightless space to achieve delicate separation of complex, nearly identical substances. A Johnson & Johnson subsidiary is working with NASA to conduct an extensive research program. Jesco von Puttakamer, a long-range planner at NASA, estimates that space manufacturing could lower the production cost of a drug that fights blood clots from its current $1,200 a dose to around $100, perhaps saving some of the 50,000 people who die of clots each year. Because cooling metals do not form convection currents in a vacuum, strong alloys can be made from metals that are grainy and crumble when combined on earth. Thus the German auto manufacturer BMW is reportedly interested in fashioning lead and aluminum alloy motor parts that would be highly heat resistant, self-lubricating and last for a half million miles of use.[73]

Expert opinion about the prospects for space manufacturing is sharply divided, reflecting the limited number of experiments that have been performed and the large uncertainties about the costs of setting up a commercially viable factory in space. The many experiments in materials processing scheduled for the shuttle-launched Spacelab in 1983 will answer some questions, while undoubtedly creating even more. Because the most critical factor in the economics of space processing will be the weight of the raw materials and the processing equipment, the first markets are likely to be for products with a high value-to-weight ratio, such as pharmaceuticals and electronic devices, followed by specialty glasses and alloys. A study for NASA by Microgravity Research Associates estimates that two or three $500-million space factories, weighing 15,000 kilograms and serviced every three months by the space shuttle, could meet half the growth in demand in semiconductors in the United States over the next decade.[74]

If these visions of space manufacturing materialize, the products will be quite expensive, limiting their use to markets such as advanced weapons systems, scientific instruments and medical devices where performance, not price, is critical. However, the most probable payoff from space materials experiments is likely to be new knowledge of how materials interact that can be applied in terrestrial manufacturing. The principal financial supporters of Spacelab, the West Germans, look to these experiments to augment their knowledge of basic materials science rather than as precursors to industrial manufacturing in space. Successful space manufacturing ventures will almost surely be carried out by Western high technology corporations, thus intensifying the gap between the North and the South.[75]

Although space manufacturing may become a modest industry some day, large-scale space industrialization is an unworkable attempt to escape from the problems of earth. Much of the recent writing about humanity's future in space has been dominated by a group of outlandish proposals for industrializing space—solar power satellites, space colonies and asteroid mining—that aim to bypass the earth's resource limits by importing energy and materials from space or exporting people from the planet. At first glance these massive undertakings have a certain logic: the earth is limited, space is infinite. Outer space does contain vastly more energy, space and materials than earth. Yet this abundance cannot be brought to bear meaningfully on the earth's problems. Based as they are on misleading ideas about the nature of humanity's earthbound resource predicament, attempts to realize these fantasies are likely to exacerbate the problems of the earth during a period of unprecedented environmental stress. Rather than signs of hope in a time of troubles they are the most grandiose hallucinations of technological civilization.

The plans for the large-scale habitation and exploitation of space have evoked extensive public enthusiasm, which reflects at least a growing realization that the present course of growth on earth cannot continue indefinitely. Long a staple of science fiction writers, space colonies were first seriously proposed in the early seventies by Dr. Gerard O'Neill, a Princeton University physicist. In *The High Frontier: Human Colonies in Space,* O'Neill detailed plans to build colonies with at first 10,000 inhabitants, and later a million. Manufactured out of materials from the earth, and then the moon, the colonies would be completely self-sufficient by harnessing the sun for all sources of energy. Space colonization is specifically promoted as a solution to overcrowding and environmental degradation on the earth. By exporting ever greater numbers of humans into these orbiting cities, the wildlife and wilderness qualities of the earth could be protected, he claimed, or perhaps even expanded.[76]

Life there is envisioned as pastoral, pollution-free, pluralistic—sort of floating garden cities. "Artists recreations" depict vast transparent domes with sunlight flowing in for plants and lighting. The reality of space habitation would be far bleaker for centuries to come. The radiation belts around the earth make these visions misleading to the point of fraud. Heavy thick metal shielding would be necessary to block the lethal quantities of cosmic and solar radiation. Life in space for the foreseeable future will be like that in a submarine, an off-shore oil platform or an Antarctic mining camp—dangerous, cramped, isolated and uneventful.[77]

Large-scale space colonization is not now—and may never be—technologically feasible. Structures of the size envisioned are thousands of times larger than anything yet built in space. If the record of similar scale-ups on earth is any guide, there will be unforeseen and perhaps insurmountable problems. Ecologist Paul Ehrlich points out that scientists have no idea how to create large, stable ecosystems of the sort that would be needed to make space colonies self-sufficient. The key to such knowledge is, of course, much more study of the ecosystems on earth, many of which are becoming less diverse and less stable. It could be many decades before scientists know enough to understand—let alone recreate—ecosystems as complex as those now being degraded.[78]

Nor are space colonies a meaningful response to the population and environmental problems of the earth. At a time when 800 million humans live in "absolute poverty" on earth, it makes little sense to think about building fabulously expensive habitats in space. Simply transporting the world's daily increase of about 200,000 people into space would consume the annual gross national product of the United States. And long before technology made space colonies feasible the current rate of population growth would have caused the earth's life support systems to collapse. More basically, human population growth is not an inevitable natural necessity requiring an ever more expensive technical response. Population stabilization is a difficult social challenge, but it is certainly less complex than the organizational and political skills needed for large-scale colonization of space. Even if feasible, the endless multiplication of human beings in these hive canisters is ultimately as pointless a vision as it is bleak.[79]

The space resource venture given the most attention as having near-term relevance to humanity is the solar power satellite, or SUNSAT, envisioned first in 1968 by Dr. Peter Glazer. A solar power satellite system would consist of three parts: an orbiting array of photovoltaic cells to collect and turn sunlight into electricity, a microwave beamer to send the power through the earth's atmosphere and fields of microwave antennae on earth to convert the beams back into usable electricity. Variant designs would use lasers in the place of micro-

"Life in space for the foreseeable future will be like that in a submarine, an off-shore oil platform or an Antarctic mining camp—dangerous, cramped, isolated and uneventful."

waves and mirrors in the place of photovoltaic cells. The system's principal appeal is its ability to collect virtually unlimited amounts of solar energy without obstruction from the atmosphere or the cycle of night and day and without polluting the earth's atmosphere with combustion by-products. During the seventies, as energy costs rose and environmental constraints became increasingly visible, SUNSAT proponents backed by large U.S. aerospace corporations created considerable public and government interests in the idea. They claimed that mass production of photovoltaic cells and new rockets many times the size of those in use today would make this scientifically feasible venture into one that was also technologically and economically viable.[80]

The construction of SUNSAT would be an undertaking of unprecedented size and cost. A 1980 NASA and U.S. Department of Energy study estimated that 60 satellites, each as big as Manhattan Island, would be needed to produce 300 gigawatts of electricity, which is roughly the current U.S. usage. The cost estimates ranged from an optimistic $1.5 trillion to a more probable $3 trillion. If two SUNSATs were built each year, one heavy-lift rocket—seven times the size of the largest rocket ever built—would have to be launched each day for 30 years to build and service them.[81]

After an initial burst of favorable press by space and energy groups, several in-depth studies of SUNSAT by panels of independent scientists and analysts concluded that the concept faced technical, economic, environmental and political hurdles that are probably insurmountable. The U.S. Office of Technology Assessment concluded that SUNSATs involved as many environmental risks as large-scale terrestrial energy systems such as coal and nuclear power—and more technological uncertainties. A 1981 study by the U.S. National Academy of Sciences went a step further, concluding that obstacles to solar power satellites were so great that no funds should be spent for its development for at least the next decade, a remarkable recommendation for a group usually highly sympathetic to research.[82]

The barriers to an economical SUNSAT are technological rather than scientific. At each step in its construction, technological barriers would have to be broken and costs would have to fall rapidly. Photovoltaic cells are made already, for example, and there may soon be cheaper ones. But scientists do not know how to make cheap ones with the ability to withstand intense radiation and cosmic ray bombardment for decades. Similarly, many major advances and scale-ups in rocket design and the handling of materials are needed to build such large structures in space. One major uncertainty in the cost of SUNSAT is labor. Preliminary evidence suggests that workers could not stay in orbit at the construction site for more than a few days without absorbing unsafe levels of radiation. Unless the health of thousands of highly trained workers is to be sacrificed on this high-energy altar, work tours would have to be short, driving up training and transportation costs.[83]

More troubling than these technical constraints are planetary scale environmental risks. Beaming trillions of watts of microwaves through the atmosphere for extended periods of time is almost certain to alter the composition of gases in unpredictable ways. Launching millions of tons of material into orbit would also place large quantities of exhaust gases into the upper atmosphere, perhaps disrupting the thin bands of volatile gases, such as the ozone layers, that play important but poorly understood roles in regulating the habitability of the earth. An operational SUNSAT system big enough to make a difference in the terrestrial energy equation would be a shot-in-the-dark experiment with the human race's one habitable environment—the earth's atmosphere.[84]

The political and institutional barriers to SUNSAT are less quantifiable but no less intractable. These large satellites, powerful sources of electromagnetic radiation, would orbit in a relatively small band of near space that already carries a great deal of satellite traffic, so other uses of space would be constrained. At the minimum, international agreements would have to ensure that energy benefits reached those who would have to forego other use of these orbits. The operation of such a system would involve worldwide resource allocation regimes far beyond those now in existence. If these massive power systems ever fell into the hands of terrorists or politically ambitious generals, or fell back into the earth's atmosphere, the results could be disastrous.[85]

One other mirage of abundance in space that has recently received attention is asteroids. A scheme to mine them to meet mineral needs

"An operational SUNSAT system would be a shot-in-the-dark experiment with the human race's one habitable environment—the earth's atmosphere."

on earth has also been proposed in some detail by former U.S. astronaut Brian O'Leary. These irregularly shaped rocks that orbit the sun, mainly between Mars and Jupiter, range in size from as small as a grain of sand to as big as the state of Texas. Although our direct knowledge of asteroids in space is limited, scientists have studied the composition of a few that have over the millenia collided with the earth. Many asteroids have nickel and iron contents several times as high as ores now mined and, unlike metallic ores on the earth, do not require energy-intensive smelting. Indeed, the world's largest and highest grade nickel deposit, in Sudbury, Ontario, is believed to be the remains of a giant asteroid that struck the planet over two billion years ago. Although there is probably enough metal in the asteroid belt to meet world needs for many centuries, getting the asteroids to the earth's surface would be costly and energy-intensive, and would risk an accidental collision and ecological disruption on a colossal scale. Long before it becomes feasible or economical to bring rare metals in from space, scientists should be able to turn the prodigiously abundant clay, silicon, aluminum, hydrocarbons and iron in the earth's crust into plastics, ceramics, specialty irons and glasses to meet global materials requirements.[86]

Whose Space?

The useful areas of near space are perhaps best understood as extensions of earth. While vast in size by terrestrial standards, near space is a limited, degradable resource. The increasingly used part of space is subject to many of the same problems afflicting the earth—resource disputes, crowding and pollution. As with any valuable resource, questions of ownership are of great political consequence.

Space is an irreducibly global entity. Because objects remain in near space only if they are rapidly orbiting the earth, the concept of national sovereignty works no better in space than it does in the ocean basins and the atmosphere, where air and water circulate indifferent to human borders. As with these other global commons, space is an arena for conflicts between the North and South. There has been an attempt to fashion a legal regime for space resource use from the "common heritage of mankind" principle, but it has not been widely accepted.

For the Soviet Union and the United States, the crowding of space poses a painful dilemma. While needing international agreement to continue to enjoy the benefits of space, the superpowers cannot set up a regime without sharing technology and reserving some orbit slots for the have-not countries. For the Third World, the growing lead of the industrial nations in exploiting this new global resource is seen as a threat to their painfully achieved and economically insecure independence.

The central agreement governing outer space is the 1967 Treaty on Principles Governing the Activities of States in the Exploration and Use of Outer Space, Including the Moon and Other Celestial Bodies, referred to as the Outer Space Treaty. This was drafted by the U.N. Committee on the Peaceful Uses of Outer Space (COPUOS), a standing U.N. body that legislates on all matters concerning space. The core doctrine of the treaty is that outer space is not subject to national appropriation. It has been ratified by 107 countries, including all those active in space.[8]

A second U.N. space treaty, a 1979 agreement known as the Moon Treaty, is not yet in effect and has been ratified by just three countries, only one of which, France, is active in space. The Moon Treaty takes the principle of non-appropriation one step further, declaring that the moon and other celestial bodies are the "common heritage of mankind." While the Outer Space Treaty was essentially a codification of the status quo in space, the Moon Treaty is a bold effort to lay down principles for activities that do not yet—and may never—exist. It has drawn considerable criticism for trying to restrict future activities without a clear knowledge of what those activities might be. In response, Third World advocates say the treaty only establishes that future lawmaking will be governed by general principles that protect their minimum interests. Prospects for U.S. ratification are dim; aerospace and mining companies are lobbying vigorously against it, claiming—probably with some exaggeration—that the treaty would effectively forbid private enterprise on the moon. Nor is the Soviet Union, which is generally opposed to the "common heritage of mankind" doctrine, likely to sign it soon. The Moon Treaty is likely to remain in legal limbo for the foreseeable future.[88]

"Some areas of the geosynchronous orbit over parts of the equator south of Europe and North America are already fairly crowded."

The Outer Space Treaty contains a provision banning the stationing of "weapons of mass destruction"—nuclear bombs—in orbit. Unfortunately, this agreement, often cited as an important arms control accomplishment, is not verifiable and bans a technology with neglible military value. Since high yield nuclear weapons are today the size of a small suitcase, hiding them aboard satellites would be simple. While the vision of nuclear weapons spinning around the world ready to descend when called is frightening, orbiting bombs are far less accurate than ballistic missiles, can only strike a narrow band of the earth beneath their orbital path, and pose nightmarish command and control problems. Despite these problems, an historian of the period notes that President Kennedy accepted the Soviet invitation to negotiate the ban against the unanimous advice of his top national security advisers—a reflection of the kind of pressure arms controllers face from military establishments eager to preserve all options.[89]

Curiously, the Outer Space Treaty does not contain a definition of outer space, leaving unanswered the question of where the atmosphere ends and space begins, or where national airspace becomes international space. This could be a considerable problem in the years ahead should some countries attempt to extend and enforce national sovereignty over very high overflights. Reconnaissance and remote sensing satellites orbiting 80 miles above the earth could be prime targets.[90]

Ownership disputes caused by crowding in space are likely to emerge first over the geosynchronous orbits, a band of space 22,500 miles above the equator. A satellite's orbit is called geosynchronous when the speed of the satellite around the earth matches the speed of the earth's rotation, so that the satellite is stationary relative to a particular place on earth. To avoid radio interference, satellites in geosynchronous orbit must be placed some distance from each other, limiting the number that can use this valuable band to 180. Some areas of the geosynchronous orbit over parts of the equator south of Europe and North America are already fairly crowded, and INTELSAT projects that slots in some areas will run out by the early nineties.[91]

Most communications and many weather satellites are in geosynchronous orbit. To achieve continuous coverage otherwise requires movable antennas and several satellites, because some are speeding behind the earth at any one time. These alternatives obviously increase severalfold the cost of a given service. Because satellites in geosynchronous orbit are so high, they can cover a large area. In fact, three communications satellites cover all the world's surface except the polar circles.

Thus far geosynchronous orbits have been taken on a first come, first served basis. Developing countries claim this discriminates against them because many of the slots will be full by the time they can make extensive use of communication satellites. The first effort to stake a claim on this valuable orbital resource was the Bogota Declaration, signed by seven equatorial countries in 1976. Since satellites in geosynchronous orbit hover over them, rather than simply flying over like other satellites, these countries declared these orbits were extensions of their territorial airspace. This claim has generally not been respected because many Third World countries would also be excluded by it and because it contradicts the "non-appropriation" principle at the heart of both the Outer Space and Moon Treaties. A more realistic system, proposed by India, would set up a licensing system for these orbital slots, awarding a certain number to each country. These permits could then be sold, enabling use of the space now but preserving minimum ownership rights for developing countries.[92]

The United States has strenuously opposed any international regime governing the allocations of geosynchronous orbits, arguing that technological innovation in transmission systems will allow a continuous expansion of the resource. Even if this "technical fix" works, an international agreement on the universal use of the more compact transmission technology will be needed, thus opening the door to a legal regime that the United States will not be able to control. Another plausible technical response to crowding is to build large platforms where many satellite functions could be combined. This, however, would also probably require international involvement. One way or another, the world seems headed toward a legal order to allocate the increasingly scarce and valuable geosynchronous orbits.[93]

Another area of conflict in space concerns the legal role of private enterprise, a struggle of increasing importance as private companies begin to build launch vehicles. The Soviets and the Third World argue that private firms can be allowed in space only as extensions of state activity, while the United States, Western Europe and Japan maintain that private enterprise under license from governments can operate in space. Developing countries fear their nascent national programs could be overshadowed by these private efforts. If, however, the basic objective of private enterprise in space—lower cost service—can be achieved, the developing countries could expand their use of space without paying the high costs of acquiring an independent space capability.[94]

The exploration of space has entailed the export of not only territorial conflict and resource disputes, but also pollution. Already between 10,000 and 15,000 objects—satellites that no longer function, empty fuel tanks and plain garbage—are orbiting the earth, mainly in the most useful areas. In addition, several million small metal fragments from explosions and satellite killer tests litter near space. These objects travel at speeds of several kilometers per second and can easily punch right through satellites or spaceships. No craft carrying astronauts has been damaged seriously yet, but some experts attribute the unexplained failure of several satellites to collision with space debris. Scientists are worried that their ability to perform sensitive experiments will be compromised by space litter. The large number of artificial objects in orbit is also of concern to military officials who increasingly question their ability to detect the difference between space clutter and a surprise missile attack. With orbital debris increasing at a rate of 11 percent a year, a panel of experts at the American Institute of Aeronautics and Astronautics recently concluded that this exotic pollution would pose "unacceptable risk" to human use of space within a decade. A full-scale war between the superpowers in space would dramatically degrade its usefulness for future generations.[95]

As on earth, pollution control in space will require a combination of regulation, technology change, legislation against some activities altogether and perhaps even special garbage collection vehicles. Outlawing explosions and killer satellite tests is the most important first step. Rockets and satellites can be redesigned to eject less material while operating and can be equipped with small booster engines to carry them into deeper space at the end of their useful lives. Specially designed space-litter collectors could pluck larger pieces of debris out of the most useful orbits. International standards in this area are woefully lacking and should be drawn up now rather than after a highly publicized disaster.[96]

The exhaust from rockets traveling through the upper atmosphere poses a potentially great, though currently insignificant, threat to the ozone layer. This form of pollution has received little attention aside from an environmental impact statement prepared by NASA in 1972 for the space shuttle. NASA concluded that 60 flights of the shuttle would deposit some 5,500 tons of hydrogen chloride into the stratosphere each year. This in turn would cause an annual decrease in the ozone layer of 0.25 percent, an amount NASA says would be undetectable against much larger, naturally occurring fluctuations in atmospheric ozone. But even this could be reduced greatly if spaceships used liquid hydrogen and oxygen fuel systems that leave only water vapor behind. The impact of rocket exhaust on the upper atmosphere, however, is likely to place constraints on any large-scale use of space.[97]

Inadvertent reentry into the atmosphere of orbiting nuclear power systems is another hazard of the space enterprise to be reckoned with. Most orbiting satellites are powered by solar energy collected by panels of photovoltaic cells. Others, however, particularly those that send out powerful radar beams, rely on nuclear power. Experts estimate that between 20 and 100 satellites now in orbit contain altogether about a ton of enriched uranium. Objects that reenter the atmosphere are normally vaporized before they reach the earth's surface, but the heavy metal shielding around nuclear reactors helps ensure that some parts will remain intact and hit the ground. The greatest danger is from the low orbiting Soviet military satellites that contain nuclear reactors. Some American military satellites are powered by radioactive isotopes, but these are much less radioactive and in higher orbits, where an inadvertent reentry is less likely. Because photovoltaic panels are easily crippled by debris, nearby explosions or lasers, the number of satellites with reactors is bound to increase as the United States and the Soviet Union harden their militarily significant satellites against attack.[98]

The potential hazards of these orbiting reactors was suddenly brought home when a Soviet ocean surveillance satellite, COSMOS 954, unexpectedly reentered the atmosphere, spreading debris over a

"For all our looking and probing of the universe, we have yet to find anywhere as habitable as the remotest, most forbidding parts of this planet."

remote part of northern Canada. Although the overall risk of such accidents is not high, several countries have called for controls on their use. Suggestions include banning them outright, requiring international notification of their launch and orbital path, and making nuclear-equipped craft move above the low orbits where inadvertent reentry is more probable.[99]

The overall record of international attempts to resolve conflict in space, as on earth, has not been particularly productive. Treaties have outlawed activities not likely to occur anyway or have avoided the difficult questions and focused instead on vague principles without clear meaning. Thus an unverifiable ban on militarily useless atomic weapons has been agreed upon while the growing problem of anti-satellite weapons remains unregulated. The principle of non-appropriation has not solved problems of conflicting use and, in its broader expression in the Moon Treaty, has done little to advance the interests of the Third World nations. The general irrelevance of the legal regime in outer space is the clear product of the unwillingness of the Soviet Union and the United States to enter into any agreement that meaningfully restricts their activities. In the years ahead the increased use of space and the growing number of participants will see either new and more meaningful agreements or conflict and collision. Unless any regulatory efforts are flexible enough to encompass evolving technologies and new, unexpected uses, however, chaos will be replaced by a straight jacket. Only with a wider sense of shared commitment and less national insecurity will the world be able to enjoy the full benefits of space.[99]

Toward an Earth-Oriented Space Program

Space's most important lesson is its reaffirmation that humanity's fate will be determined on earth. Colonization of space in the distant future is something scientists can speculate and science fiction writers can imagine. But for all our looking and probing of the universe, we have yet to find any place as habitable as the remotest, most forbidding parts of this planet. Space exploration has taught us just how rare and precious the earth is.

Knowing how little there is in space gives us all the more reason to preserve the fullness of life on earth. The loss of a species takes on a cosmic importance when there is nothing else as complex or fragile within trillions of miles. Within the broader horizons of the human spirit opened by space, the most important priorities are, then, preventing the mass detonation of nuclear arsenals on earth and attending to neglected earthkeeping tasks such as reforestation, population growth and the protection of genetic diversity.

The second lesson of the quarter-century in space is that near space—despite all its physical dissimilarities from the earth—can best be understood as an extension of the human world. Near space now routinely affects daily life and, just like the earth, is the site of ideological competitions, warped spending priorities, resource disputes and relative indifference to the interests of future generations. It is time to abandon the view that space is some remote frontier where survival of the fittest is the law. People must start behaving in space the way they would in their own backyards: they must take care that nothing they do harms their nearby home and attend to the concerns of their neighbors.

Unfortunately, much of the current thinking about space is dominated by the lawlessness and escapism of the frontier mentality. The superpowers have a first come, first served attitude toward a global medium that is no more their own than the air they breathe or the ocean straits through which they demand right of passage. Those with a positive space program succumb to the old frontier illusion that an Eden of abundance and harmony awaits us in space. The urge to pick up and move to a new land when things start getting bad in the old country has taken on a new high-technology character.

The advocates of war in space are similarly escapist. Recognizing that war on earth is too horrible to contemplate, they want to ship it out into space where it will not hurt anyone, a position that has a surprisingly diverse group of supporters. This image of a space war being like an eighteenth-century battle, when armies fought other armies while civilian life went on largely unaffected, does not bear up to examination. The military space activities of the superpowers are too interconnected with the command, targeting and warning sys-

"The ability to make nuclear weapons more accurate and more deadly is the biggest—but least noticed—space news of the last decade."

tems of strategic nuclear weapons on earth: a war in one realm would almost surely become a war in the other.[101]

The space age is the result of a revolution in rocket propulsion. The most significant impact of this new technology has been in the military sphere. Rocket propulsion technology has been most extensively employed to make the intercontinental artillery that is now poised to destroy much of civilization on short notice. The ability to warn about, control and command nuclear arsenals from space, and to make weapons more accurate and more deadly, is the biggest—but least noticed—space news of the last decade. Space technology is the other leg of the unprecedented and continuing multiplication of destructive potential known as the nuclear revolution. It would be wonderful if a sudden technical fix like a space-based anti-ballistic missile system eliminated the threat of nuclear war, but arms control agreements verified from space are a far more foolproof and vastly less expensive path to national survival and peace. Although it can help make war infinitely destructive, space technology in the form of surveillance satellites could serve as the keystone for its abolition. Extending that capability and integrating it into arms control agreements is the greatest foreseeable contribution space can make to human welfare.

Looking beyond the superpowers, the most disturbing trend is the worldwide spread of rocket technology in a context of national rivalry rather than global cooperation. The spread of space launch technology has not been subject to as much attention as the spread of nuclear technology, despite its equally ambiguous and potentially destructive character. As with nuclear weapons, the superpowers cannot expect the Third World to forego a technology with civilian as well as military uses at the very time they are rushing to militarize space.

Although the developing countries cannot force the superpowers to cooperate, they can avoid letting the introduction of militarily ambiguous launch capability heighten their own regional tensions. Both as pioneering efforts in fashioning new global institutions and as measures to live peaceably with their neighbors, Third World cooperation in space has merit. A Chinese-Indian-Pakistani program would make particular sense because these emerging space powers share insecure borders and long traditions of hostility; none can afford a buildup of arms but all could use space to help solve pressing resource problems. Such a program could be modeled after the cooperative success of the European Space Agency. Often openly and rightfully critical of the North for its failure to cooperate, the countries of the South have a chance to avoid imitating the nationalism and militarism that flaw the world's space accomplishments to date.

The overwhelmingly military character of the space venture has been curiously neglected by the many space visionaries who have played such a decisive role in opening the high frontier. Like many people with a new or powerful idea they seem indifferent to the means required to achieve their end. This particular blindness was perhaps best captured in satirist Tom Lehrer's proposal to add to the title of space pioneer Wernher von Braun's autobiography, *I Aim for the Sky,* "but sometimes I hit London." In contrast to the outspoken protest and personal anguish of many atomic scientists, there has been little resistance to space militarization among the leading space developers. Until recently the military's need to use space has immeasurably helped the space advocates reach their goal—to explore and exploit space. But today this relation has been reversed, putting the militarization of space squarely in the path of peaceful space development.

In August 1982 the nations of the world will meet under U.N. auspices for the first time since 1968 to confer on the future of outer space. When representatives from all parts of the planet reach Vienna for UNISPACE 82, they will find the militarization of space curiously absent from the official agenda. But no meaningful discussion of the conference's official topic of the peaceful uses of outer space can occur unless the participants tackle the biggest threat—the progressive militarization of space by the Soviet Union and the United States. The superpowers have kept those military activities off the agenda. A concerted effort by the rest of the world should be made to ensure that this and future meetings put the threat of space war where it belongs—at the top of the program.

The international scientific community and the countries just now beginning to use space are logical leaders in the global space demilitarization effort. Just as the consequences of a "limited nuclear war" on earth could be catastrophic for everyone, so too would a "limited war" in space have a far-reaching impact. Even if it did not spread to earth, the explosion of only a few nuclear weapons in orbit would disable billions of dollars worth of operating civilian satellites as well as further degrade the space environment. The effects of the escalating militarization of space on the international scientific community are even more immediate and disturbing: cooperative agreements cancelled, missions eliminated altogether and technology monopolized. The willingness of leading scientists to speak out on the dangers of fallout from atmospheric nuclear testing in the fifties led to the Limited Test Ban Treaty of 1963. A similar stand today against space war could be effective.[102]

However bleak the prognosis, no assessment of space can ignore the sense of excitement and purpose that has accompanied the opening of the high frontier. Using space for the good of science and the earth will launch a burst of beneficial space activities—a global earth environment monitoring network, a renewed commitment to explore deep space and new communications, broadcast and remote sensing satellites. Just because war led people into space does not mean that removing warlike activities will sap the urge to explore or use it. This was, of course, the vision animating John Kennedy's challenge of a race to the moon—an effort to channel the warlike urge into peaceful competition. Now if people can take one more step and "race" toward a goal with practical value, the excitement of the early space age can be captured.

Today's frontier is not technological. In many critical areas of life and death, our technology has outstripped the competence of our political institutions. Humanity's greatest challenges now are averting war and stabilizing the physical basis of human life on earth. Engineering triumphs in space will not help in these battles. But reversing the spread of weapons into space, making space a stronger pillar of international peacekeeping, designing new institutions of global scientific cooperation, and ensuring a permanently habitable earth would be to conquer humanity's real frontiers.

The question of how much the world should be spending on space activities is not an easy one to answer. Activities such as communications satellites that yield a profit above their cost are the exception, not the rule. For most space activities, intangible values like knowledge, improved nuclear missile accuracy or prestige must somehow be weighed against terrestrial goals. It does little good to point out, as space advocates do, that people spend more on cosmetics than on space. Or to argue, as critics do, that less is spent curing deadly tropical diseases than exploring Mars. That resource allocation patterns are so warped provides little insight into the value of space activities. Whether we spend more or less than we now do exploring the universe and our place in it, a more earth-oriented space program will yield vastly greater benefits.

Renewed emphasis on peacekeeping, scientific research and monitoring the earth's life-support systems will drive countries naturally toward greater cooperation. Of all the activities in space thus far, science and communications are the areas where cooperation has been greatest. Studying the earth's planetary aspects and sending probes to other planets has required and evoked a global exchange of information. These missions have been one of the few areas in the world arena when Americans and Russians interact non-belligerently. Communications satellites would hardly be possible without cooperation, since communicating is an inherently reciprocal process. And only when a LANDSAT remote sensing system is built cooperatively (that all countries have access to and benefit from) will it be possible to convince some that they are not being spied upon. Building a global war monitoring system is the most challenging cooperative space venture—and the one with the highest payoff. Renewed interaction across the range of other space sciences and applications should make such a peacekeeping satellite network much easier to establish.

Fortunately, new cooperative space ventures do not need new institutions—just an imitation or resuscitation of old ones. A global LANDSAT system would follow in the footsteps of INTELSAT; a U.S. shuttle-Soviet Salyut linkup in the eighties could imitate the Apollo-Soyuz union of the seventies; and a Third World launch program could follow the lines of the European Space Agency. These cooperative ventures will not eliminate the roots of war or seriously compromise the sovereignty of nations. But they will provide badly needed bridges of peace and lessons in constructive coexistence.

The space venture of the last quarter-century has been one of humanity's most exciting technological accomplishments. Its greatest payoff—increased awareness of our place in the cosmos—can be the cornerstone of a new set of global priorities. Now it remains to put space to work so that we can firmly establish that there is indeed intelligent life on earth.

Notes

1. Since 1958, the United States Government has spent about $150 billion on military and civilian space activities. Hard data on USSR spending is lacking, but the Soviet Union has spent at least as much and possibly half again as much more, bringing the superpower's space expenditures to date to well over $300 billion (in current dollars). Charles Sheldon and Marcia Smith, "Space Activities of the United States, Soviet Union and Other Launching Countries/Organizations" Congressional Research Service, Washington, D.C., February 1982.

2. Joseph H. Jackson, *Pictorial Guide to the Planets* (New York: Thomas Y. Crowell, 1973).

3. Richard S. Lewis, *From Vinland to Mars* (New York: Quadrangle/The New York Times Book Co., 1976); William Bainbridge, *The Spaceflight Revolution* (New York: John Wiley & Sons, 1976).

4. Frederick Ordway, III, and Mitchell Sharpe, *The Rocket Team* (New York: Thomas Y. Crowell, 1979); Wernher von Braun and Frederick I. Ordway, III, *History of Rocketry and Space Travel* (New York: Thomas Y. Crowell, 1975).

5. Richard Hutton, *The Cosmic Chase* (New York: New American Library, 1981); Nicolas Daniloff, *The Kremlin and the Cosmos* (New York: Alfred K. Knopf, 1972).

6. National Aeronautics and Space Administration (NASA); Office of Public Affairs, "Satellite Situation Report," Goddard Space Flight Center, Greenbelt, Md., April 30, 1982.

7. James E. Oberg, *Red Star in Orbit* (New York: Random House, 1981); Mose L. Harvey, "Preeminence in Space: Still a Critical National Issue," *Orbis*, Winter 1969; Vernon Van Dyke, *Pride and Power: The Rationale of the Space Program* (Urbana: University of Illinois Press, 1964).

8. Charles S. Sheldon, "United States and Soviet Progress in Space: Summary Data and a Forward Look," Congressional Research Service, Washington, D.C., 1980; William H. Shauer, *The Politics of Space: A Comparison of the Soviet and American Space Programs* (New York: Holmes & Meier, 1976).

9. Nicolas Mateesco Matte, *Space Policy and Programmes Today and Tomorrow: The Vanishing Doupole* (Toronto: Carswell Co., Ltd., 1980).

10. James Oberg, "The Soviet Aim: A Permanent Base in Space," *New Scientist,* October 1, 1981; "National Paper: USSR," prepared for the Second U.N. Conference on the Exploration and Peaceful Uses of Outer Space, Vienna, August 9-21, 1982; B. Belitsky, "Soviet Manned Space Flight 20 Years On," *Spaceflight,* May 1981.

11. John Logsdon, *The Decision to Go to the Moon, Project Apollo and the National Interest* (Chicago: University of Chicago Press, 1970); "The Moon Landing and Its Aftermath," *Michigan Quarterly Review,* Spring 1979; John Noble Wilford, "The Spinoff from Space," *New York Times Magazine,* January 29, 1978; Courtney G. Brooks, James M. Brimwood, and Lloyd S. Swenson, Jr., *Chariots for Apollo: A History of Manned Lunar Spacecraft* (Washington, D.C.: NASA, 1978).

12. Erlend A. Kennan and Edmund H. Harvey, *Mission to the Moon, A Critical Examination of NASA and the Space Program* (New York: William Morrow & Co., 1969); Amitai Etzioni, *The Moon-Doggle, Domestic and International Implications of the Space Race* (Garden City, N.Y.: Doubleday and Co., Inc., 1964); Jerry Grey, *Enterprise* (New York: William Morrow & Co., 1979).

13. Jerry Grey, "Implications of the Shuttle: Our Business in Space," *Technology Review,* October 1981; John Noble Wilford, "Revival for One-Shot Rockets," *New York Times,* November 2, 1982; John Noble Wilford, "NASA Raises Rates for Shuttle Customers," *New York Times,* June 16, 1982; Klaus Heiss, "New Economic Structures for Space in the Eighties," *Astronautics & Aeronautics,* January 1981.

14. John Noble Wilford, "U.S. Ends Most Exploration of Space with Russia," *New York Times,* June 6, 1982; Peter Jankowitsch, *International Cooperation in Outer Space,* Occasional Paper 11 (Muscatine, Iowa: Stanley Foundation, 1976); Edward C. and Linda N. Ezell, *The Partnership: A History of the Apollo-Soyuz Test Project* (Washington, D.C.: NASA Scientific and Technical Information Office, 1978); Valnora Leister, "Export Controls of Space Technology: An Overlooked Barrier to International Cooperation," in *Proceedings of the Twenty-Third Colloquium on the Law of Outer Space* (New York: American Institute of Aeronautics and Astronautics, 1981).

15. James Oberg, "Hitch Up with a Red Star," *Omni,* March 1982.

16. European Space Agency, *Europe's Place in Space* (Paris: 1981); Walter McDougall, "The Scramble for Space," *Wilson Quarterly,* Autumn 1980; Peter Marsh, "What Should Europe Do in Space?," *New Scientist,* January 29, 1981.

17. Ministry of International Trade and Industry, *Report of the Deliberation Council on Basic Problems in the Space Industry,* Tokyo, April 20, 1981; Neville Kidger, "Japanese Space Plans for 1982-1985," *Spaceflight,* May 1982; Masayoshi Kanabayashi, "Japan Sets Mini-Entry in Space Race," *Wall Street Journal,* July 18, 1981.

18. Theo Pirard, "Space Systems Operated and Prepared by Developing Countries," *Spaceflight,* May 1981; David Richie, "Dragon in the Sky: China's Space Program," *Technology Review,* October 1981; John Noble Wilford, "Chinese Advance in Bold Space Program," *New York Times,* March 11, 1980; H. P. Mama, "India's Space Program: Across the Board on a Shoe String," *Interavia,* January 1980; Stuart Auerbach, "Peaceful Indian Program Has Military Potential," *Washington Post,* December 4, 1981; Azim Kidwai, "Muslim Middle East Moves to Exploit Space," *Pakistan & Gulf Economist,* May 15-21, 1982; Salim Alvi, "Space Technology in South Asia: New Vistas Opening," *Pakistan & Gulf Economist,* May 15-21, 1982; S. M. Fazal interview with Salim Mehmud, "The Future of Space Technology in Pakistan," *Pakistan & Gulf Economist,* May 15-21, 1982.

19. Judith Miller, "Industry Enters a Twilight Zone," *New York Times,* September 13, 1981; Judith Miller, "U.S. Reportedly Concerned Over Uses of Commercially Produced Rockets," *New York Times,* September 12, 1981; "A Bargain-Basement Challenge to NASA," *Business Week,* June 22, 1981; M. Mitchell Waldrop, "Firm Offers to Finance a Space Shuttle," *Science,* February 12, 1982.

20. Office of Technology Assessment, U.S. Congress, *Civilian Space Policy and Applications* (Washington, D.C.: June 1982).

21. Denver Research Institute, *Space Benefits: The Secondary Application of Aerospace Technology in Other Sectors of the Economy* (Denver, Colo.: University of Denver for NASA, January 1981); Alex Roland, "Returns to Earth," *Wilson Quarterly,* Autumn 1980; Frederick Ordway, Carsbie C. Adams, and Mitchell R. Sharpe, *Dividends from Space* (New York: Cromwell, 1972).

22. Matte, *Space Policy and Programmes Today and Tomorrow;* Congressional Research Service, *Soviet Space Programs 1971-75,* Report for the Senate Committee on Aeronautical and Space Sciences (Washington, D.C.: August 1976); "Background for International Programs," in Office of Technology Assessment, *Civilian Space Policy.*

23. Sheldon, "United States and Soviet Progress in Space"; J. Preston Layton, "Military Space in Transition," *Astronautics & Aeronautics,* October 1981; U.S. Department of Defense, "Soviet Military Power," Washington, D.C., 1982; John Noble Wilford, "About a Fourth of NASA Budget for Research is Said to Be Military," *New York Times,* May 3, 1982.

24. Nicholas Johnson, "The Military and Civilian Salyut Space Programmes," *Spaceflight,* August/September 1979; R. Jeffrey Smith, "Military Plans for Shuttle Stir Concern," *Science,* May 1, 1981; Gerald L. Borrowman, "The Military Role in the Shuttle," *Spaceflight,* May 1982.

25. Philip J. Klass, *Secret Sentries in Space* (New York: Random House, 1971); Ted Greenwood, "Reconnaissance and Arm Control," *Scientific American,* February 1973.

26. Les Aspin, "The Verification of the SALT II Agreement," *Scientific American,* February 1979; Stuart A. Cohen, "The Evolution of Soviet Views on SALT Verification: Implications for the Future," in William Potter, ed., *Verification and SALT: The Challenge of Strategic Deception* (Boulder, Colo.: Westview Press, 1980).

27. Robert A. Manning and Stephen Talbot, "Carter's Nuclear Deal with South Africa," *Inquiry,* October 30, 1978; "U.S. Monitoring Chinese-Vietnamese War Through Space Photos and Radio Intercepts," *New York Times,* March 1, 1979; Phil Gailey, "Senator Jackson Sees Pressing Need for Expanded U.S.-Soviet Hot Line," *New York Times,* April 26, 1982; Jozef Goldblat, *Agreements for Arms Control: A Critical Survey* (Oelgeschlager, Gun & Hain, Inc.: Cambridge, Mass., 1982).

28. Curtis Peebles, "The Guardians," *Spaceflight,* November 1978; Anthony Kendon, "U.S. Reconnaissance Satellite Programs," *Spaceflight,* July 1978; Leslie Gelb, "Keeping an Eye on Russia," *New York Times Magazine,* November 29, 1981.

29. Preparatory Committee for the Second Special Session of the General Assembly Devoted to Disarmament, "Study on the Implications of Establishing an International Satellite Monitoring Agency," United Nations, August 1981; Government of France, "On the Establishment of an International Satellite Monitoring Agency," report submitted to the United Nations, June 11, 1979; Howard and Harriet Kurtz, *A Global Information Complex* (Washington, D.C.: War Control Planners, Inc., 1979); Ram S. Jakhu and Riccardo Trecroce, "International Satellite Monitoring for Disarmament and Development," *Annals of Air & Space Law,* 509 (1980); "Symposium Report," *Pugwash Newsletter,* April 1980.

30. Preparatory Committee for Session on Disarmament, "Study on the Implications of International Monitoring Agency."

31. Thomas H. Karas, *Implications of Space Technology for Strategic Nuclear Competition,* Occasional Paper 25 (Muscatine, Iowa: Stanley Foundation, July 1981); Deborah Shapley, "Technology Creep and the Arms Race: ICBM Problem a Sleeper," *Science,* September 22, 1978.

32. Richard L. Garwin, "Trends in the Technological Development of Nuclear Weapons Systems," unpublished; R. P. Denaro, "Navstar: The All-Purpose Satellite," *IEEE Spectrum,* Vol. 18, No. 5, 1981.

33. Herbert Scoville, Jr., and Kosta Tsipis, *Can Space Remain a Peaceful Environment?,* Occasional Paper 18 (Muscatine, Iowa: Stanley Foundation, July 1978).

34. "The New Military Race in Space," *Business Week,* June 4, 1979; Leonard David, "The Military Uses of Outer Space," presented to the 23rd Annual Meeting of the American Astronautical Society, San Francisco, October 18-20, 1977; Stockholm International Peace Research Institute, *Outer Space—Battlefield of the Future?* (London: Taylor & Francis, Ltd., 1978); Lt. Gen. Daniel O. Graham (Ret.), *High Frontier: A New National Strategy* (Washington, D.C.: The Heritage Foundation, 1982).

35. Mark C. Hallam, *Star Wars Revisited: A Look at the Arms Race in Outer Space* (Washington, D.C.: SANE, 1978); Robert C. Aldridge, "Who Will Shoot First in Outer Space?," *The Nation,* March 25, 1978; William Broad, "Nuclear Pulse (II) Ensuring Delivery of the Doomsday Signal," *Science,* June 7, 1981.

36. Michael Getler, "Soviet Missile Test: Scenario for War," *Washington Post,* June 21, 1982; "Satellite Killers," *Aviation Week & Space Technology,* June 21, 1976; Thomas O'Toole, "Russians Resume Testing of Satellite Designed to Hunt and Destroy Others," *Washington Post,* April 19, 1980.

37. Robert Toth, "War in Space," *Science 80*, September/October 1980.

38. Richard Holloran, "Pentagon Plans System to Fight Soviet Satellites," *New York Times*, June 6, 1982; Michael A. G. Michaud, "The Anti-Satellite Program a Threat to Space Humanization?," in *The Space Humanization Series* (Arlington, Va.: Institute for the Social Science Study of Space, 1979).

39. "Technology Seen Hindering Efforts to Limit Arms Race," *Washington Post*, November 29, 1976; "Soviets See Shuttle as Killer Satellite," *Aviation Week & Space Technology*, April 17, 1978; "Chinese Space Gains Hamper Antisatellite Limitation Treaty," *Aviation Week & Space Technology*, July 9, 1979.

40. Donald Hafner, "Averting a Brobdingnagian Skeet Shoot: Arms-Control Measures for Antisatellite Weapons," *International Security*, Winter 1980/81; Stephen M. Meyer, "Antisatellite Weapons and Arms Control: Incentives and Disincentives from Soviet and American Perspectives," *International Journal*, Summer 1981; David Andelman, "Space Wars," *Foreign Policy*, Fall 1981; Hans A. Bethe and Kurt Gottfried, "The 5-Year Plan—A Loser 2 Ways," *New York Times*, June 10, 1982.

41. William J. Beane, "The High-Energy Laser: Strategic Policy Implications," *Strategic Review*, Winter 1977; Clarence A. Robinson, Jr., "Laser Technology Demonstration Proposal," *Aviation Week & Space Technology*, February 16, 1981; Richard Burt, "Experts Believe Laser Weapons Could Transform Warfare in 80's," *New York Times*, February 10, 1980; Jeff Hecht, "Lasers Make Ready for War," *New Scientist*, June 10, 1982; David Binder, "U.S. & Soviet Reported Trying to Build Missile-Destroying Beam," *New York Times*, February 5, 1977.

42. David Richie, "Laser Rattling in Outer Space," *Inquiry*, September 1, 1980.

43. William Broad, "A Fatal Flaw in the Concept of Space War," *Science*, March 12, 1982; Kosta Tsipis, "Laser Weapons," *Scientific American*, December 1981; G. Bekefi *et al.*, *Particle Beam Weapons* (Cambridge, Mass.: Program in Science and Technology for International Security at MIT, 1978).

44. John Noble Wilford, "Riding High," *Wilson Quarterly*, Autumn 1980.

45. John Noble Wilford, "A Rich Era in the Study of Planets Draws to a Close," *New York Times*, August 25, 1981; Christine Sutton, "A New Decade of Science and Astronomy in Space," *New Scientist*, January 8, 1981.

46. Peter Marsh, "Robots Leap Into the Space Age," *New Scientist,* April 23, 1981; Christine Sutton, "Machines Uncover the Secrets of Space," *New Scientist,* October 1, 1981; Ewald Heer, "New Luster for Space Robots and Automation," *Astronautics & Aeronautics,* September 1978.

47. Eugene H. Levy, "On a National Policy for Space Science: The Advisory Basis," presented to the 148th National Meeting of the American Association for the Advancement of Science, Washington, D.C., January 6, 1982; M. Mitchell Waldrop, "Planetary Science in Extremis," *Science,* December 18, 1981.

48. John Noble Wilford, "Space Telescope Holds NASA's Hopes for Grand Discoveries in Universe," *New York Times,* January 5, 1982; John N. Bahcall and Lyman Spitzer, Jr., "The Space Telescope," *Scientific American,* July 1982.

49. Bruce Murray, Michael C. Malin, and Ronald Greeley, *Earthlike Planets: Surfaces of Mercury, Venus, Earth, Moon, Mars* (San Francisco: W. H. Freeman & Co., 1981); Carl Sagan, *The Cosmic Connection* (New York: Dell, 1973); Bruce Murray, "Space Exploration—Is It Worth the Cost?," *Planetary Report,* Spring 1980.

50. James E. Lovelock, *GAIA: A New Look at Life on Earth* (Oxford: Oxford University Press, 1979); Roger Bingham, "The Maverick and the Earth Goddess," *Science 81,* December 1981.

51. Space Science Board, *A Strategy for Earth Science from Space in the 1980's, Part I: Solid Earth and Oceans* (Washington, D.C.: National Academy Press, 1981); P. A. Hanle and V. D. Chamberlain, eds., *Space Science Comes of Age* (Washington, D.C.: Smithsonian Institution Press; 1981); "World Climate, the Oceans, and Early Indications of Climatic Change," in Office of Technology Assessment, *Civilian Space Policy.*

52. John Noble Wilford, *The Mapmakers* (New York: Alfred A. Knopf, 1981); "Oceanography from Space," *Oceanus,* Fall 1981; Thoms O'Toole, "Satellite Records a Steady Decline in the Sun's Radiant Energy," *Washington Post,* April 20, 1982; Edward A. Flinn, "Application of Space Technology to Geodynamics," *Science,* July 3, 1981; several articles on SEASAT, *Science,* June 29, 1979; "Nimbus Traces Ozone Changes," *Aviation Week & Space Technology,* March 24, 1980.

53. W. J. Hussey and E. L. Heacock, "The Economic Benefits of Environmental Satellites," National Oceanic and Atmospheric Administration, U.S. Department of Commerce, 1978; "Typhoons, Worst Foe of Farmers, Bind 11 Nations," *Christian Science Monitor,* April 19, 1982; H. W. Lorber and R. H. Drake, "The Economic Benefits of Space Development," Los Alamos Laboratory, Los Alamos, New Mex., March 1982.

54. Conference Secretariat, "Relevance of Space Activities to Monitoring of Earth Resources and the Environment," prepared for the Second U.N. Conference on the Exploration and Peaceful Uses of Outer Space, Vienna, August 9-21, 1982; European Space Agency, *Earth Observation from Space & Management of Planetary Resources*, Conference Proceedings, Toulouse, France, March 6-11, 1978; Paul Kinnucan, "Earth-Scanning Satellites Lead Resource Hunt," *High Technology*, March/April 1982; Committee on the Peaceful Uses of Outer Space, "The Detection and Monitoring of Pollution of the Environment by Means of Remote Sensing from Outer Space," United Nations, New York, September 23, 1977.

55. "Crop Forecasting from Space: Toward a Global Food Watch," *Science*, May 2, 1975; "Satellite Keeps Tabs on Forests," *Journal of Commerce*, May 6, 1982; Klaus Heiss, "United States Benefits of Improved Worldwide Wheat Crop Information from a LANDSAT System Overview," in European Space Agency, *Earth Observation from Space;* Marvin E. Bauer, "The Role of Remote Sensing in Measuring Crop Production," Inter-American Development Bank, Washington, D.C., May 30, 1978; General Accounting Office, "Crop Forecasting by Satellite: Progress and Problems," Washington, D.C., April 7, 1978.

56. The Geosat Committee, Inc., "Satellite Remote Sensing Data—An Unrealized Potential for the Earth Science Community," San Francisco, 1977; Thomas Wayland, "Remote Sensing Continues to Open New Horizons for Ore Seekers," *Engineering & Mining Journal*, July 1976; "U.S. Probe Uncovers Copper," *Washington Post*, February 22, 1975; William L. Smith, ed., *Remote Sensing Applications for Mineral Exploration* (Stroudsburg, Penn.: Dowden, Hutchinson & Ross, Inc., year of publication unknown).

57. Ad Hoc Committee on Remote Sensing for Development, *Resource Sensing from Space: Prospects for Developing Countries* (Washington, D.C.: National Academy of Sciences, 1977); Charles K. Paul and Adolpho C. Mascarenhas, "Remote Sensing in Development," *Science*, October 9, 1981; United Nations, *The Application of Space Technology to Development* (New York: 1973); Philip M. Fearnside, "Deforestation in the Brazilian Amazon: How Fast Is It Occurring?," *Interciencia*, March/April 1982.

58. Stephanie Yanchinski, "Thorny Questions Over Remote Sensing," *New Scientist*, April 17, 1980; D. M. Polter, "Remote Sensing and State Sovereignty," *Journal of Space Law*, Fall 1976; N. M. Matte and H. DeSaussure, eds., *Legal Implications of Remote Sensing From Outer Space* (Amsterdam: A. W. Sijthoff-Leyden, 1976).

59. Office of Technology Assessment, *Civilian Space Policy,* Centre National d'Etudes Spatiales, "SPOT Satellite-Based Remote Sensing," city of publication unknown, 1982; M. Mitchell Waldrop, "Imaging the Earth (II): The Politics of LANDSAT," *Science,* April 2, 1982; M. Mitchell Waldrop, "Imaging the Earth (I): The Troubled First Decade of LANDSAT," *Science,* March 26, 1982.

60. American Society of International Law, *A Global Satellite Observation System for Earth Resources: Problems and Prospects* (Washington, D.C.: National Academy of Sciences, 1977); Charles K. Paul and Adolpho C. Mascarenhas, "Remote Sensing in Development," *Science,* October 9, 1981; United Nations, *The Application of Space Technology to Development* (New York: 1973); Philip M. Fearnside, "Deforestation in the Brazilian Amazon: How Fast Is It Occurring?," *Interciencia,* March/April 1982.

61. David Mathisen, "The Unfulfilled Promises of Remote Sensing," *New Engineer,* June/July 1978.

62. Arthur C. Clarke, "Extraterrestrial Relays: Can Rocket Stations Give Worldwide Radio Coverage?," *Wireless World,* October 1945.

63. S. Astrain, "Telecommunications and the Economic Impact of Communications Satellites," presented to the 31st Congress of the International Astronautical Federation, Tokyo, September 21-28, 1980; Burton I. Edelson and Robert S. Cooper, "Business Use of Satellite Communications," *Science,* February 12, 1982.

64. S. Astrain, *INTELSAT Annual Report 1980* (Washington, D.C.: International Telecommunications Satellite Organization, 1980); Lester R. Brown, *World Without Borders* (New York: Random House, 1972); J. N. Pelton and M. S. Snow, eds., *Economic and Policy Problems in Satellite Communications* (New York: Praeger Publishers, 1977).

65. Joe Pelton, *Global Talk,* (Alphen aan den Rijn, Netherlands: Sijlhoff and Noordhoff, 1981); Heather E. Hudson *et al., The Role of Telecommunications in Socio-Economic Development* (Geneva: International Telecommunications Union, 1979).

66. B. I. Edelson and R. D. Briskman, "Space Applications: The Communications Outlook," presented to the 1981 Annual Meeting of the American Institute of Aeronautics and Astronautics, Long Beach, Calif., May 12-14, 1981; James Martin, *Future Developments in Telecommunications* (Englewood Cliffs, N.J.: Prentice-Hall, Inc., 1977); John McLucas, "Global Cooperation in Satellite Communications in a Decade of Policy Divergence," presented to the International Aerospace Symposium, Paris, June 2, 1981.

67. Roy Rowan, *The Four Days of Mayaguez* (New York: W. W. Norton & Co., 1975); Zbignew Brzezinski, "The Failed Mission," *New York Times Magazine,* April 18, 1982.

68. "Pinpoint Surveying: Can Oilmen Hitch Their Wagons to Navstar?," *Energy Daily,* March 22, 1982; Duncan Campbell, "Astro-Navigation Goes Artificial," *New Scientist,* March 9, 1978; Geoff Richards, "Navstar—A Complete Global Naviation System," *Spaceflight,* January 1980.

69. E. Pervyshin, "The Curtain Goes Up on the Olympics," *Telecommunications,* October 1979.

70. Wilson P. Dizard, "The U.S. Position: DBS and Free Flow," *Journal of Communication,* Spring 1980; Laura Landro, "Backyard Satellite Dishes Spread But Stir Fight with Pay-TV Firms," *Wall Street Journal,* April 2, 1982; Barry Newman, "European States Face Problem of Controlling Their Neighbors' TV," *Wall Street Journal,* March 22, 1982; U.S. House of Representatives, Subcommittee on Telecommunications, Consumer Protection, and Finance, Committee on Energy and Commerce, *Satellite Communication/Direct Broadcast Satellites,* Hearings, December 15, 1981; Les Brown, "Are the Networks Dinosaurs," *Channels,* June/July 1982.

71. Bella Mody, "The Indian Satellite Instructional TV Experiment: Its Origins, Organizations, Messages, and Effects," Institute for Communication Research, Stanford University, Calif., unpublished, 1979; Bella Mody "Lessons from the Indian Satellite Experiment," Institute for Communication Research, Stanford University, Calif., unpublished, 1980; Robert Hornik, "Mass Media Use and the Revolution of Rising Frustrations," *Communication Research,* Vol. 9, No. 4, 1977; Binod Agarwal, "Satellite Instructional Television Experiment: Television Comes to Village—An Evaluation of SITE," Indian Space Research Organization, New Delhi, October 1978.

72. Henry Cooper, "Shuttle I & II," *New Yorker,* February 9, and 16, 1981.

73. J. E. Bishop, "NASA, Two Firms Plan to Create Exotic Drugs in First 'Space Factory'," *Wall Street Journal,* September 23, 1981; Jesco von Puttkamer, "Industrialization of Space," *Futurist,* June 1979.

74. National Research Council, *Materials Processing in Space* (Washington, D.C.: National Academy of Sciences, 1978); G. Harry Stine, *The Third Industrial Revolution* (New York: Ace, 1979); Donald M. Waltz, "The Promise of the Space Factory," *Technology Review,* May 1977.

75. National Research Council, "Practical Applications of Space Systems," National Academy of Sciences, Washington, D.C., 1975.

76. Gerard K. O'Neill, *The High Frontier: Human Colonies in Space* (New York: William Morrow & Co., 1977); J. P. Vajk, "The Impact of Space Colonization on World Dynamics," *Technological Forecasting and Social Change,* January 1976; Gerard K. O'Neill, "Space Colonies and Energy Supply to the Earth," *Science,* December 5, 1975; Eric Orexler, "The Space Colonies Idea, 1969-1977," in Stewart Brand, ed., *Space Colonies,* (New York: Penguin, 1977).

77. T. A. Heppenheimer, *Colonies in Space* (Harrisburg, Pa.: Stakpole Books, 1977); "Statement of John Holt," in Brand, *Space Colonies.*

78. "Statement of Paul and Anne Ehrlich," in Brand, *Space Colonies.*

79. World Bank, *World Development Report* (Washington, D.C.: 1981); Lester R. Brown, *Building a Sustainable Society* (New York: W. W. Norton & Co., 1981).

80. P. E. Glaser, "Power From the Sun: Its Future," *Science,* November 22, 1968; U.S. House of Representatives, Committee on Science and Technology, *Solar Satellite Power System Concepts,* Hearings, February 20, 1976.

81. Solar Power Systems Concept Development and Evaluation Program, *Program Assessment Report, Statement of Findings* (Washington, D.C.: Department of Energy/NASA, November 1980).

82. National Science Foundation, *Electric Power from Orbit: A Critique of a Satellite Power System* (Washington, D.C.: 1981); Office of Technology Assessment, U.S. Congress, *Solar Power Satellites* (Washington, D.C.: 1981); R. A. Herendeen, T. Kary, and J. Rebitzer, "Energy Analysis of the Solar Power Satellite," *Science,* August 3, 1979.

83. David Rust, "Solar Flares, Proton Showers, and the Space Shuttle," *Science,* May 28, 1982; Office of Technology Assessment, *Solar Power Satellite.*

84. W. E. Gordon and L. M. Duncan, "SPS Impacts on the Upper Atmosphere," *Astronautics & Aeronautics,* July/August 1980; Department of Energy, *Environmental Assessment for the Satellite Power System—Concept Development and Evaluation Program, Atmospheric Effects* (Washington, D.C.: November 1980).

85. Peter Vajh, "On the Military Implications of Satellite Power Systems," *Lincoln Proceedings,* April 1980; Paul L. Csonka, "Space Colonization: An Invitation to Disaster?," *The Futurist,* October 1977; Michael Ozeroff, "SPS Military Implications," Department of Energy/NASA, Washington, D.C., October 1978.

86. Walter Sullivan, "Three New 'Apollo Objects' Sighted Near Earth," *New York Times,* April 13, 1982; Brian O'Leary, *The Fertile Stars* (New York: Everest House, 1981); Brian O'Leary, "Mining the Apollo and Amor Asteroids," *Science,* July 22, 1977.

87. Carl Christon, "The Growth of Space Law," *Astronautics & Aeronautics,* December 1981; "The International Legal Regime of Outer Space," in Office of Technology Assessment, *Civilian Space Policy; Space Law, Selected Basic Documents,* 2nd edition (Washington, D.C.: Government Printing Office, 1978).

88. Office of Technology Assessment, "Agreement Governing the Activities of States on the Moon and Other Celestial Bodies," Washington, D.C., 1980; "Statement of Leigh S. Ratiner, on Behalf of the L-5 Society," in the U.S. House of Representatives, Subcommittee on Space Science and Applications, Committee on Science and Technology, *International Space Activities,* Hearings, September 6, 1979; Delbert Smith, "The Moon Treaty and Free Enterprise," *Astronautics & Aeronautics,* January 1980; Nancy Griffin, "Americans and the Moon Treaty," *Journal of Air Law & Commerce,* Spring 1981; Emilio Jaksetic, "Peaceful Uses of Outer Space: Soviet Views," *American University Law Review,* Vol. 28, Summer 1979.

89. Raymond L. Garthoff, "Banning the Bomb in Outer Space," *International Security,* Winter 1980/81; Goldblat, *Agreements for Arms Control.*

90. G. P. Zhukov, "Delimitation of Outer Space," in *Proceedings of the Twenty-Third Colloquium on the Law of Outer Space.*

91. Howard Hupe, "Carving Up the Geosynchronous Orbit," *Astronautics & Aeronautics,* February 1978; Maureen Williams, "The Geostationary Orbit: A Limited Natural Resource from Outer Space," *International Relations,* November 1979.

92. Harvey J. Levin, "Outer and Spectrum Resource Strategies: Third World Demands," *Telecommunications Policy,* June 1981; Office of Technology Assessment, U.S. Congress, *Radiofrequency Use and Management: Impacts from the World Administrative Radio Conference of 1979* (Washington, D.C.: January 1982).

93. U.S. Government, "A Description of Alternative Methods for Planning the Fixed-Satellite Service," submitted to International Telecommunication Union, March 17, 1981; Conference Secretariat, "Efficient Use of the Geostrategy Orbit," prepared for the Second U.N. Conference on the Exploration and Peaceful Uses of Outer Space, Vienna, August 9-21, 1982; Kim G. Gibbons, "Orbital Saturation: The Necessity for International Regulation of Geosynchronous Orbits," *California Western International Law Journal,* Winter 1979.

94. A. M. Dula, "Regulation of Private Commercial Space Activities," in *Proceedings of the Twenty-Fourth Colloquium on the Law of Outer Space* (New York: American Institute of Aeronautics and Astronautics, 1982).

95. Technical Committee, "Space Debris," American Institute of Aeronautics and Astronautics, New York, July 1981; Bernard Lovell, "The Pollution of Space," *Bulletin of the Atomic Scientists,* December 1968; S. E. Doyle, "Reentering Space Objects: Facts & Fiction," *Journal of Space Law,* Fall 1978.

96. "Trash in Orbit: Growing Hazard for Astronauts," *U.S. News & World Report,* April 19, 1982.

97. Lydia Dotto and Harold Schiff, *The Ozone Wars* (Garden City, N.Y.: Doubleday & Co., 1978); NASA, "Environmental Statement for the Space Shuttle Program, Final Statement," Washington, D.C., July 1972.

98. Fernando Lay, "Nuclear Technology in Outer Space," *Bulletin of the Atomic Scientists,* September 1979.

99. Thomas O'Toole, "Soviet Satellite Burns Up Over Canada," *Washington Post,* January 25, 1978; U.S. Government, "Use of Nuclear Power Sources in Outer Space," submitted to U.N. Committee on the Peaceful Uses of Outer Space, New York, February 1979.

100. Eilene Galloway, "Consensus Decisionmaking of UNCOPUOS," *Journal of Space Law,* Vol. 7, Spring 1979; Oliver William Hennigan, Jr., "Near Earth Orbital Space: Implications for American Foreign Policy," *Fletcher Forum,* Vol. 4, January 1980, Richard Berryman and Richard Schifter, "The New International Regulation: A Global Straitjacket," *AEI Journal of Government and Society,* September/October 1981; S. B. Rosenfield, "A Moon Treaty? Yes, But Why Now?," in *Proceedings of the Twenty-Third Colloquium on the Law of Outer Space.*

101. Robert Salkeld, *War and Space* (Englewood Cliffs, N.J.: Prentice-Hall, 1970); Stewart Brand, "War in Space: Good," *New Scientist*, May 28, 1981.

102. Robert A. Divine, *Blowing on the Wind: The Nuclear Test Ban Debate 1954-1960* (New York: Oxford University Press, 1978).

DANIEL DEUDNEY is a Senior Researcher with Worldwatch Institute, where his research deals with energy, resources and global commons. He has studied Philosophy and Political Science at Yale University, and Science, Technology and Public Policy at the George Washington University.